Time Series Indexing

Implement iSAX in Python to index time series with confidence

Mihalis Tsoukalos

‹packt›

BIRMINGHAM—MUMBAI

Time Series Indexing

Associate Group Product Manager: Kaustubh Manglurkar

Publishing Product Manager: Dhruv Jagdish Kataria

Senior Editor: Tiksha Lad

Technical Editor: Devanshi Ayare

Copy Editor: Safis Editing

Book Project Manager: Hemangi Lotlikar

Proofreader: Safis Editing

Indexer: Manju Arasan

Production Designer: Joshua Misquitta

Marketing Coordinator: Vinishka Kalra

First published: June 2023

Production reference: 1120623

Published by Packt Publishing Ltd.
Livery Place
35 Livery Street
Birmingham
B3 2PB, UK.

ISBN 978-1-83882-195-1

www.packtpub.com

To my niece, Marietta.

– Mihalis Tsoukalos

Contributors

About the author

Mihalis Tsoukalos holds a BSc in mathematics from the University of Patras and an MSc in IT from University College London, UK. His books *Go Systems Programming* and *Mastering Go* have become must-reads for Unix and Linux systems professionals. He enjoys writing technical articles and has written for Sys Admin, Mactech, C/C++ Users Journal, USENIX ;login:, Linux Journal, Linux User and Developer, Linux Format, and Linux Voice. His research interests include time series data mining, time series indexing, machine learning, and databases.

Mihalis is also a photographer.

As writing a book is a team effort, I would like to thank the people at Packt Publishing for helping me write this book. Especially, I would like to thank Tiksha Abhimanyu Lad, Hemangi Lotlikar, and Dhruv J. Kataria.

Many thanks should go to the people at the University of the Peloponnese, at the Department of Informatics and Telecommunications, for introducing me to the time series and time series data mining research fields.

Last, I would like to thank Devanshu Tayal, the technical reviewer, for his comments and his suggestions.

About the reviewer

Devanshu Tayal is a data science enthusiast with experience in travel and banking. He has a master's degree from **Birla Institute of Technology and Science (BITS)** Pilani and has also studied at the IK Gujral Punjab Technical University, in mechanical engineering. In his spare time, Devanshu enjoys researching new applications for data science, playing music, and badminton. Diversity and inclusion, Python, algorithms, data structures, machine learning, **Natural Language Processing (NLP)**, Tableau, Power BI, data visualization, and AI are some of his other interests.

Table of Contents

3

iSAX – The Required Theory   65

4

iSAX – The Implementation   97

5

Joining and Comparing iSAX Indexes   131

6

Visualizing iSAX Indexes   159

7

Using iSAX to Approximate MPdist   183

8

Conclusions and Next Steps   213

Preface

The title of the book you are reading is *Time Series Indexing*, which should hint at its contents.

The time series index discussed and explored in this book is called iSAX. iSAX is considered one of the best indexes for time series, which is the main reason for choosing it. Besides implementing iSAX and the SAX representation as Python 3 packages, this book shows how to work with time series at the subsequence level and understand the information presented in academic research papers.

But the book does not stop here as it presents Python scripts for getting to know your time series data better and code for visualizing time series data and iSAX indexes to better understand the data as well as the structure of a particular iSAX index.

Who this book is for

This book is for developers, researchers, and university students of any level that want to work with time series at the subsequent level and use a modern time series index in the process.

Although the presented code is in Python 3, once you understand the ideas and the concepts behind the code, the packages and scripts can easily be ported to any other modern programming language, such as Rust, Swift, Go, C, Kotlin, Java, JavaScript, and so on.

What this book covers

Chapter 1, An Introduction to Time Series and the Required Python Knowledge, is all about the fundamentals that you need to know to follow this book, including the importance of time series and how to set up a proper Python environment to run the code of the book and experiment with time series.

Chapter 2, Implementing SAX, explains SAX and the SAX representation and presents Python code for computing the SAX representation of a time series or a subsequence. It also presents Python scripts that calculate statistical quantities that can give a higher overview of a time series and plot histograms of your time series data.

Chapter 3, iSAX – The Required Theory, presents the theory behind the construction and the use of the iSAX index and shows how to manually construct a small iSAX index step by step using lots of visualizations.

Chapter 4, iSAX - The Implementation, is about developing a Python package for creating iSAX indexes that fit in memory and presents Python scripts that put that Python package into action.

Chapter 5, Joining and Comparing iSAX Indexes, shows how to use iSAX indexes created by the isax package and how to join and compare them. At the end of the chapter, the subject of testing Python code is discussed. Last, we show how to write some simple tests for the isax package.

Chapter 6, Visualizing iSAX Indexes, is all about visualizing iSAX indexes using various types of visualizations using the JavaScript programming language and the JSON format.

Chapter 7, Using iSAX to Approximate MPdist, is about using iSAX indexes to approximately compute the Matrix Profile vectors and the MPdist distance between two time series.

Chapter 8, Conclusions and Next Steps, gives you directions on what and where to look next if you are really into time series or databases by proposing classical books and research papers to study.

To get the most out of this book

This book requires a UNIX machine with a relatively recent Python 3 installation and the ability to install Python packages locally. This includes any machine running recent versions of macOS and Linux. All the code has been tested on a Microsoft Windows machine.

We propose that you use software for Python package, dependency, and environment management to have a stable Python 3 environment. We use Anaconda, but any similar tool is going to work fine.

Last, if you really want to make the best use of the book, then you need to experiment as much as you can with the presented Python code, create your own iSAX indexes and visualizations, and maybe port the code into a different programming language.

Download the example code files

You can download the example code files for this book from GitHub at https://github.com/ PacktPublishing/Time-Series-Indexing. If there's an update to the code, it will be updated in the GitHub repository.

We also have other code bundles from our rich catalog of books and videos available at https:// github.com/PacktPublishing/. Check them out!

Download the color images

We also provide a PDF file that has color images of the screenshots and diagrams used in this book. You can download it here: https://packt.link/Pzq1j

Conventions used

There are a number of text conventions used throughout this book.

`Code in text`: Indicates code words in text, database table names, folder names, filenames, file extensions, pathnames, dummy URLs, user input, and Twitter handles. Here is an example: "We are going to perform our experiments using the following sliding window sizes: 16, 256, 1024, 4096, and 16384."

A block of code is set as follows:

```
def query(ISAX, q):
    global totalQueries
    totalQueries = totalQueries + 1
    Accesses = 0


    # Create TS Node
```

When we wish to draw your attention to a particular part of a code block, the relevant lines or items are set in bold:

```
    # Query iSAX for TS1
    for idx in range(0, len(ts1)-windowSize+1):
        currentQuery = ts1[idx:idx+windowSize]
        found, ac = query(i1, currentQuery)
        if found == False:
            print("This cannot be happening!")
            return
```

Any command-line input or output is written as follows:

```
$ ./accessSplit.py -s 8 -c 32 -t 500 -w 16384 500k.gz
Max Cardinality: 32 Segments: 8 Sliding Window: 16384 Threshold: 500
Default Promotion: False
OVERFLOW: 01111_10000_10000_01111_10000_01111_10000_01111
Number of splits: 6996
Number of subsequence accesses: 19201125
```

> **Tips or important notes**
> Appear like this.

Get in touch

Feedback from our readers is always welcome.

General feedback: If you have questions about any aspect of this book, email us at `customercare@packtpub.com` and mention the book title in the subject of your message.

Errata: Although we have taken every care to ensure the accuracy of our content, mistakes do happen. If you have found a mistake in this book, we would be grateful if you would report this to us. Please visit `www.packtpub.com/support/errata` and fill in the form.

Piracy: If you come across any illegal copies of our works in any form on the internet, we would be grateful if you would provide us with the location address or website name. Please contact us at `copyright@packtpub.com` with a link to the material.

If you are interested in becoming an author: If there is a topic that you have expertise in and you are interested in either writing or contributing to a book, please visit `authors.packtpub.com`.

Share Your Thoughts

Once you've read *Time Series Indexing*, we'd love to hear your thoughts! Scan the QR code below to go straight to the Amazon review page for this book and share your feedback.

`https://packt.link/r/1838821953`

Your review is important to us and the tech community and will help us make sure we're delivering excellent quality content.

Download a free PDF copy of this book

Thanks for purchasing this book!

Do you like to read on the go but are unable to carry your print books everywhere?

Is your eBook purchase not compatible with the device of your choice?

Don't worry, now with every Packt book you get a DRM-free PDF version of that book at no cost.

Read anywhere, any place, on any device. Search, copy, and paste code from your favorite technical books directly into your application.

The perks don't stop there, you can get exclusive access to discounts, newsletters, and great free content in your inbox daily

Follow these simple steps to get the benefits:

1. Scan the QR code or visit the link below

https://packt.link/free-ebook/9781838821951

2. Submit your proof of purchase
3. That's it! We'll send your free PDF and other benefits to your email directly

1

An Introduction to Time Series and the Required Python Knowledge

This is the first chapter of the book that you are reading. Although first chapters usually contain basic information that you might want to skip, this is not such a chapter. It teaches you the basics of time series and indexing, as well as how to set up a proper Python environment, which is going to be used during the development of the code of this book. You might need to refer to it while you are reading other chapters, which is a good thing! So, let us get started!

In this chapter, we are going to cover the following main topics:

- Understanding time series
- What is an index and why do we need indexing?
- The Python knowledge that we are going to need
- Reading time series from disk
- Visualizing time series
- Working with the Matrix Profile
- Exploring the MPdist distance

Technical requirements

In order to follow this chapter, which is the foundation of the entire book, you need to have a recent Python 3 version installed on your computer and be able to install any other required software on your own. We are not going to teach you how to install a Python 3 package, but we are going to tell you which packages you should install and the commands that we have used to do so. Similarly, we are not going to explain the process of installing new software on your machines, but we are going to tell you the command or commands we have used to install a given software on our machines.

The GitHub repository of the book can be found at `https://github.com/PacktPublishing/Time-Series-Indexing`. The code for each chapter is in its own directory. Therefore, the code for *Chapter 1* can be found inside the `ch01` folder. You can download the entire repository on your computer using `git(1)`, or you can access the files via the GitHub user interface.

You can download the entire code of this book, including the code in the `ch01` folder, using `git(1)` as follows:

```
git clone git@github.com:PacktPublishing/Time-Series-Indexing.git
```

As the repository name is long and the local directory is named after the repository name, you can execute the previous command as follows to shorten the folder name:

```
git clone git@github.com:PacktPublishing/Time-Series-Indexing.git tsi
```

This is going to put the contents of the repository in a directory named `tsi`. Both ways are valid – do what is best for you.

The code of the book is now on your local machine. However, you are going to need to have some Python packages installed for most of the code to run – we are going to discuss the required Python packages later on in this chapter.

> **Disclaimer**
>
> This code of the book was written and tested on Arch Linux and macOS Ventura machines. Even though the book is Unix-oriented, there exist similar commands that can be executed on a Microsoft Windows machine that should not be so difficult to find and execute. What is important is the presented code, understanding the code and the logic behind it, and being able to execute and make changes to it on your own. If this information helps you, I am mainly using Microsoft Visual Studio Code to write code on both macOS and Linux.

Understanding time series

A time series is a set of data. Keep in mind that a time series does not have to contain time or date data in it – time and date data usually come in the form of *timestamps*. So, a time series might contain timestamps, but usually, it does not. In fact, most of the time series in this book do not contain timestamps. In practice, what we really need is ordered data – this is what makes a bunch of values a time series.

Strictly speaking, a time series (T) of size n is an ordered list of data points: $T = \{t_0, t_1, t_2, \dots t_{n-1}\}$. Data points can be timestamped and store a single value, a set of values, or a list of values. The index of a time series might begin with 1 instead of 0 – in this case, $T = \{t_1, t_2, t_3, \dots t_n\}$. What is truly important here is that the length of the time series is n in both cases. So, each element has an index value associated with it, which replaces the need for a timestamp. Time series in this book are going to use index values to distinguish their elements. The following ordered list can be considered a time series – `{1, -2, -3, 4, 5, 1, 2, 0.23, 4.3}`. It contains nine elements. The first element is `1` and the last element is `4.3`. If the index of the first element is `0`, then the index of the last element would be `8`, whereas if the index of the first element is `1`, then the index of the last element is going to be `9`. Time series can contain the same value multiple times.

> **An alternative definition of time series**
>
> A time series is a collection of observations made sequentially in time. Many types of observations are not true time series but can be transformed into time series.

Figure 1.1 shows a graphical representation of a time series with 1,000 elements – even with a small time series such as the one presented here, it is difficult to search for a specific subsequence or value. As we will discuss later on, this is why indexing is important.

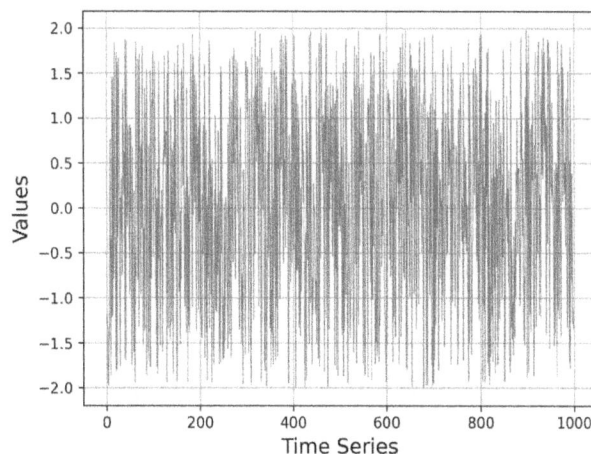

Figure 1.1 – Visualizing a time series

In the *Visualizing time series* section, we will learn how to visualize a time series in Python.

The next subsection tells us where we can find time series data.

Time series are everywhere

You might now ask where we can find time series. The answer is simple: time series are everywhere! From medical data to positional data and from software and hardware metrics to financial information and stock prices! Successfully using them allows us to find answers to questions we might have, such as which stock to sell or which hard disk is going to fail.

Let us look at some definitions that we need to know to understand the concepts better.

Essential definitions

In this subsection, we are going to learn some core definitions related to time series:

- The *length* of a time series or a subsequence is the number of elements found in the time series or the subsequence.

- A *subsequence s* of size w of a time series T is a sublist of T, with consecutive elements of length w.

- A *sliding window* of size w decomposes a time series into subsequences of size w. The sliding window separates a time series into multiple subsequences, with a length equal to the sliding window value. Given a time series with length n and a sliding window of size w, the total number of subsequences of size w is equal to $n-w+1$.

Let us now give you an example. Imagine having the following time series, T: {0, 1, 2, 3, 4, 5, 6}. Given a sliding window of size *5*, T can be separated into the following subsequences:

- `{0, 1, 2, 3, 4}`
- `{1, 2, 3, 4, 5}`
- `{2, 3, 4, 5, 6}`

So, we have three subsequences in total, each having a length of 5. As this is a tedious process, we are going to learn how to tell a computer to do that for us in this chapter.

The next subsection briefly discusses the subject of time series data mining.

Time series data mining

Data mining is the study of collecting, cleaning up, processing, analyzing, and understanding data. Data mining is a large subject. In fact, data mining is an area of computer science with its own subtopics and areas. The most important areas of data mining are the following:

- **Classification**: This is the process of determining the class label of an element given a set of predefined class labels

- **Clustering**: This is the process of grouping data into sets so that group members are similar to each other, based on a given criterion, which is usually a distance function

- **Outlier detection**: This is the process of finding an observation that differs from other ones enough to raise suspicions that it was created by a different process

Time series data mining, as the term suggests, is the data mining of time series. The main difference between regular data mining and time series data mining is that in time series, data comes sorted by time. Therefore, you cannot arrange time series data on your own. Although time gives context, what is important is the actual value.

Apart from time, time series data can also be characterized by longitude and latitude values (*spatial data*). We are not going to deal with spatial data in this book.

Having time series data is good, but it might be useless if we cannot compare this data. The next subsection shows some popular techniques and algorithms for comparing time series.

Comparing time series

To compare anything, we need a metric. We can compare numerical values because the values are the metrics. But how do we compare time series? This is an active research subject that does not have a definitive answer.

Before you continue reading the remaining chapter, take some time and try to think whether you can compare time series with a different number of elements. Is that even possible? *Write down your thoughts* before you continue reading and find out the answer.

Well, it turns out that you can compare time series with a different number of elements. However, not all metric functions support that functionality.

> **Writing and reading**
> Reading any worthwhile book or research paper is good and allows you to learn new things and keep your mind active. However, to test your knowledge and organize your thoughts, you need to write them down! I do that all the time. After all, this is how this book was created!

The Euclidean distance

The Euclidean distance is a way of finding out how close or how far apart a couple of time series are. Put simply, the Euclidean distance measures the shortest path between two multidimensional points. A time series or a subsequence with more than one element is a multidimensional point.

The Euclidean distance **prioritizes time** – it compares data points that appear at the same time and ignores anything else. Therefore, if two time series only match at different times, they are considered dissimilar. Finally, the Euclidean distance works with data of many dimensions – in this book, we use data of one dimension only. Do not confuse **multidimensional points** with **multidimensional data**. Multidimensional data contains multidimensional points. The time series of this book contains data with one dimension only (single values). However, we can consider a time series or subsequence as a *multi-dimensional point*.

The formula to calculate the Euclidean distance of two multidimensional points can be described as follows. Given a point p = $(p_1, p_2, ..., p_n)$ and a point q = $(q_1, q_2, ..., q_n)$, the Euclidean distance is the square root of the sum of all $(p_i - q_i)^2$ values:

Let us now present some examples by calculating the Euclidean distance of two subsequence pairs. The first pair is p = $\{1, 2, 3\}$ and q = $\{0, 2, 2\}$. So, first, we find all $(p_i - q_i)^2$ values:

- $(1 - 0)^2 = 1$
- $(2 - 2)^2 = 0$
- $(3 - 2)^2 = 1$

Then, we add the results: $1 + 0 + 1 = 2$.

Last, we find the square root of the result, which is approximately equal to 1.414213.

Now, imagine having the following two time series or subsequences – p = $\{1, 2, -1, -3\}$ and q = $\{-3, 1, 2, -1\}$. Although the time series have the same elements, these elements are in a different order. Their Euclidean distance can be calculated as before. First, we find all $(p_i - q_i)^2$ values:

- $[1 - (-3)]^2 = 4^2 = 16$
- $(2 - 1)^2 = 1^2 = 1$
- $(-1 - 2)^2 = (-3)^2 = 9$
- $[(-3) - (-1)]^2 = (-2)^2 = 4$

Therefore, the Euclidean distance is equal to the square root of $(16+1+9+4) = 30$, which approximately is equal to 5.4472.

A major drawback of the Euclidean distance is that it requires the two time series to be of the same length. Although there exist techniques to overcome that limitation, this is still an issue. One of the techniques involves using extrapolation to make the length of the smaller time series equal to the length of the bigger one.

Going forward, we are not going to calculate Euclidean distances manually, as *NumPy* offers a better way of doing so – this is illustrated in ed.py:

```
#!/usr/bin/env python3

import numpy as np
import sys

def euclidean(a, b):
    return np.linalg.norm(a-b)

def main():
    ta = np.array([1, 2, 3])
    tb = np.array([0, 2, 2])

    if len(ta) != len(tb):
        print("Time series should have the same length!")
        print(len(ta), len(tb))
        sys.exit()

    ed = euclidean(ta, tb)
    print("Euclidean distance:", ed)

if __name__ == '__main__':
    main()
```

The euclidean() function takes two NumPy arrays as input and returns their Euclidean distance as output, using np.linalg.norm(). This works because the Euclidean distance is the l2 norm and the default value of the ord parameter in numpy.linalg.norm() is 2, which is the reason for not specifically defining it. You do not need to remember that; just use the euclidean() function when needed.

The two time series are hardcoded in the script. Running ed.py generates the following output:

```
$ ./ed.py
Euclidean distance: 1.4142135623730951
```

The Chebyshev distance

The Chebyshev distance has a totally different logic than the Euclidean distance. This does not make it superior or inferior to the Euclidean distance, just different. If you do not know what to use, go with the Euclidean distance.

So, the Chebyshev distance between two multidimensional points is equal to the greatest distance of all $|p_i - q_i|$ values. The $||$ symbol is the *absolute value* of a quantity. Put simply, the *absolute value* of a quantity is equal to the value without the plus or minus sign.

Let us now present some examples by calculating the Chebyshev distance of two subsequence pairs. The first pair is {1, 2, 3} and {0, 2, 2}. Now, let us find the distances between the pairs:

- $|1 - 0| = 1$
- $|2 - 2| = 0$
- $|3 - 2| = 1$

So, the maximum of 1, 0, and 1 is equal to 1, which is the Chebyshev distance.

The second pair is {1, 2, -1, -3} and {-3, 1, 2, -1}. As before, we find the distances between the pairs of the points at the same position (same index):

- $|1 - (-3)| = 4$
- $|2 - 1| = 1$
- $|(-1) - 2| = 3$
- $|(-3) - (-1)| = 2$

So, the maximum of 4, 1, 3, and 2 is equal to 4, which is the Chebyshev distance of the aforementioned pair.

Later in this chapter, we are going to learn about a more sophisticated distance function, which is called *MPdist*.

Now that we know how to compare time series and subsequences, it is time to discuss indexes and indexing. Keep in mind that we cannot create an index without being able to compare its data, which includes time series data.

What is an index and why do we need indexing?

Can you imagine searching for a surname in an unsorted list of names? Can you imagine looking for a book in a library that does not sort its books based on book subject (the Dewey system) and then book title and author surname? I cannot! Both examples showcase a naïve but efficient indexing scheme. The more complex the data, the more sophisticated the index should be in order to perform quick searches and maybe updates on the data.

Figure 1.2 shows the visualization of a really small **iSAX** index – in reality, as a time series can be really huge, iSAX indexes tend to be much bigger and more complex.

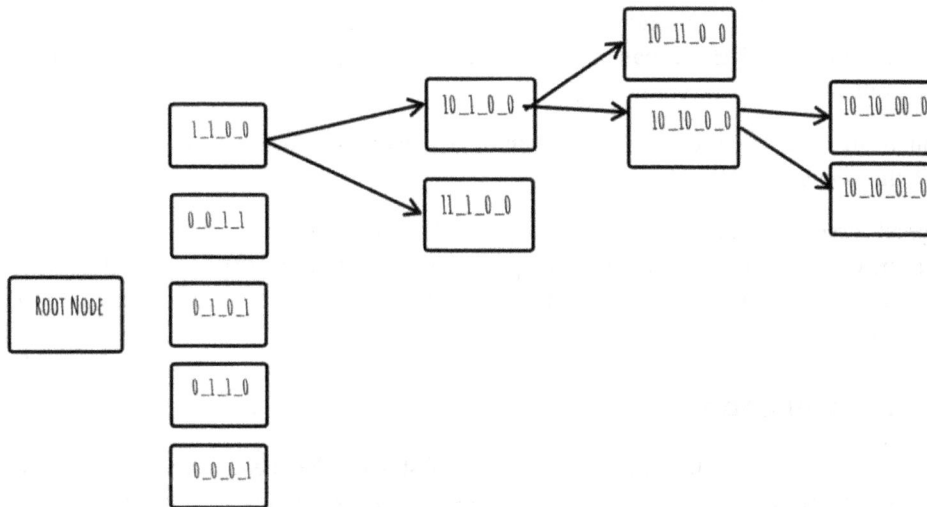

Figure 1.2 – A small iSAX index

Do not try to understand the iSAX index or the titles of the nodes at this point. Everything is going to become clearer in *Chapter 2* and *Chapter 3*. For now, keep in mind that the titles of the nodes are *SAX words* and that there exist two kinds of nodes on an iSAX index – *internal nodes* and *terminal nodes* (leaf nodes). Everything about the iSAX index and its connection with SAX words will become clear in *Chapter 3*.

In the next section, we will begin working with Python and set up our environment.

The Python knowledge that we are going to need

All the presented code in this book is written in Python. Therefore, in this section, we are going to present the required Python knowledge for you to follow this book better. However, do not expect to learn the basics of Python here – more appropriate books exist for that purpose.

What about other programming languages?

Once you learn and understand the presented theory, the Python code of this book can be easily translated into any other modern programming language, such as Swift, Java, C, C++, Ruby, Kotlin, Go, Rust, or JavaScript.

You might have compatibility issues with the used Python packages if you keep updating them for no particular reason. As a rule of thumb, I would suggest that throughout this book, you should use the same package versions, provided that they work well with each other. There exist two main ways to achieve that. You can stop upgrading your Python installation once you find the versions that work, or you can use a Python package manager such as Anaconda or pyenv. In this book, we are going to use Anaconda.

It does not matter what you use, as long as you know how to operate your tools and have a stable and reliable Python environment to work with.

I hope we all agree that the most important property of any code is correctness. However, after we have working code, we might need to optimize it, but we cannot optimize code if we do not know whether it runs slowly or not. So, the next section shows you how to calculate the time it takes Python code to execute.

Timing Python code

There are times when we need to know how slow or how fast our code is being executed because some operations might take hours or even days. This section presents a simple technique to calculate the time it takes a block of code to run.

The timing.py script shows a technique to time Python code – this might come in really handy when you want to learn how much time it takes for a process to finish. The source code of timing.py is the following:

```
#!/usr/bin/env python3

import time

start_time = time.time()
for i in range(5):
    time.sleep(1)

print("--- %.5f seconds ---" % (time.time() - start_time))
```

We use time.time() to initiate the beginning of the timing and the same statement to declare the end of the timing. The difference between these two statements is the desired result. You can also keep that difference in a separate variable.

The program executes time.sleep(1) five times, which means the total time should be pretty close to 5 seconds. Running timing.py generates the following kind of output:

```
$ ./timing.py
--- 5.01916 seconds ---
```

> **About the Python scripts**
>
> In this book, we are mostly going to show you the Python scripts in full without omitting any statements. Although this adds some extra lines, it helps you understand the functionality of the Python scripts by looking at their `import` statements before reading the actual Python code.

The next subsection is about the Anaconda software, which is used to create Python environments.

An introduction to Anaconda

Anaconda is a product for package, dependency, and environment management. Although Anaconda is a commercial product, there exists an Individual Edition for solo practitioners, students, and researchers. What Anaconda does is create a controlled environment where you can define the version of Python as well as the versions of the packages you want to use. Additionally, you can create multiple environments and switch between them.

You do not have to use Anaconda if you do not want to – however, if you are using Python 3 and you do not want to become overwhelmed by the details of Python 3 package versions, incompatibilities, and dependencies, then you should give Anaconda a try. The reason for needing package and environment management software is because some Python packages are very picky about the Python version used. Put simply, Anaconda makes sure that your Python 3 environment is not going to change and gives you the capability to transfer your Python 3 environments to multiple machines. The Anaconda command line tool is called `conda`.

Installing Anaconda

Anaconda is a huge piece of software, as it contains lots of packages and utilities. There exist multiple ways to install Anaconda, which mainly depend on your development environment.

On a macOS Ventura machine, we can install Anaconda using Homebrew as follows:

```
$ brew install anaconda
```

On an Arch Linux machine, Anaconda can be installed as follows:

```
$ pacman -S anaconda
```

We will not discuss the Anaconda installation in any more detail. The installation process is straightforward and contains lots of information. The single most important task to do is to include the Anaconda utilities on your `PATH` environment variable, in order for them to be accessible from everywhere on your UNIX shell. This also depends on the UNIX shell you are using – I use `zsh` with the Oh My Zsh extensions on both my Linux and macOS machines, but your environment might most likely vary.

If you choose to use Anaconda to work with this book, please make sure that you can access the `conda` binary and that you can enable and disable Anaconda on your machine at will – you might not need to use Anaconda all the time.

On my macOS Ventura machine, I can disable Anaconda as follows:

```
$ conda deactivate
```

I can also enable Anaconda as follows:

```
$ source /opt/homebrew/anaconda3/bin/activate base
```

You should replace base with the desired Anaconda environment.

The previous command depends on the path where Anaconda was installed. Therefore, on my Arch Linux machine, I should execute the following command instead:

```
$ source /opt/anaconda/bin/activate base
```

You should modify the previous command to fit your Anaconda installation.

When a new Anaconda version is available, you can update to the latest version by executing the next command:

```
$ conda update -n base -c defaults conda
```

Creating a new Anaconda environment

The single most important thing to decide when creating a new Anaconda environment is the selection of the Python 3 version. In order to create a new Anaconda environment named TSI that uses Python 3.8.5, you should run the following command:

```
$ conda create  --name TSI python=3.8.5
```

In order to activate this environment, run conda activate TSI. The python3 --version command shows the Python version included in a given Anaconda environment.

You can list all existing Anaconda environments using the conda info --envs command (the * character shows the active one):

```
$ conda info --envs
# conda environments:
#
TSI                      /home/mtsouk/.conda/envs/TSI
base                     /opt/anaconda
```

Changing to a different environment

This subsection presents the conda command used to change from one environment to another. Changing to a different environment is as simple as activating a different environment using conda activate environment_name.

Installing a Python package

Although you can still use `pip3` to install Python packages, the best way to install a Python package in an Anaconda environment is with the `conda install` command. Keep in mind that the `conda install` command cannot install all packages – in this case, use `pip3` instead.

Listing all installed packages

The `conda list` command gives you a full list of installed Python packages under a given Anaconda environment. As the list is pretty long, we will show you just a part of it:

```
$ conda list
# packages in environment at /home/mtsouk/.conda/envs/TSI:
#
# Name                    Version                   Build
python                    3.8.5                 h7579374_1
readline                  8.2                   h5eee18b_0
numpy                     1.23.5                    pypi_0
pandas                    1.5.2                     pypi_0
```

Deleting an existing environment

You can delete an existing Anaconda environment *that is not active* using the `conda env remove --name ENVIRONMENT` command. Its execution is illustrated in the following output when deleting an environment named `mtsouk`:

```
$ conda env remove --name mtsouk

Remove all packages in environment /home/mtsouk/.conda/envs/mtsouk:
```

This discussion of Python environments, package versions, and package incompatibilities culminates here. From now on, let us assume that we have a stable Python environment where we can use existing Python packages, develop new Python packages, and run Python scripts without any issues. The next subsection lists the Python packages that we need to install.

The required Python packages

Here is a list of the required Python packages along with an explanation of the use of each one of them:

- `NumPy`: This is the standard Python package for array computing.
- `Pandas`: This package offers data structures for data analysis, time series, and statistics, including functions to read data files from disk.
- `SciPy`: This package offers fundamental functions for scientific computing in Python.

- Matplotlib: This is the most popular Python package for scientific plotting.

- Stumpy: This is a powerful package for time series analysis and time series data mining. You do not have to install it right away, as it is not needed for the development of the iSAX index.

These are the basic packages that you need to install in a fresh Python environment. Python will automatically install any package dependencies.

Setting up our environment

In this subsection, we are going to set up our Anaconda environment. As explained earlier, this is not required to follow this book, but it will save you from Python package incompatibilities that might come up when upgrading Python and Python packages. We will execute the following commands in the TSI Anaconda environment, and then we are done:

```
(TSI) $ conda install numpy
(TSI) $ conda install pandas
(TSI) $ conda install scipy
(TSI) $ conda install matplotlib
(TSI) $ conda install stumpy
```

Printing package versions

In this subsection, we present a Python script that just loads the desired packages and prints their versions on screen.

The code of load_packages.py is the following:

```
#!/usr/bin/env python3

import pandas as pd
import argparse
import stumpy
import numpy as np
import scipy
import matplotlib

def main():
    print("scipy version:", scipy.__version__)
    print("numpy version:", np.__version__)
print("stumpy version:", stumpy.__version__)
print("matplotlib version:", matplotlib.__version__)
print("argparse version:", argparse.__version__)
print("pandas version:", pd.__version__)

if __name__ == '__main__':
    main()
```

Running `load_packages.py` on one of my UNIX machines prints the following information:

```
$ chmod 755 ./load_packages.py
$ ./load_packages.py
scipy version: 1.9.2
numpy version: 1.23.4
stumpy version: 1.11.1
matplotlib version: 3.6.2
argparse version: 1.1
pandas version: 1.5.0
```

The first command is needed to make the Python script executable and is required for all Python scripts presented in this book, provided that they begin with the `#!/usr/bin/env python3` statement. If they do not begin with that statement, you can execute them using `python3 <script_name>` without the need to change their permissions. You can learn more about the `chmod(1)` command by running `man chmod`. From now on, we will assume that you know this information and will not present any more `chmod` commands and instructions. Your output might be a little different, but that is OK, as packages get updated.

Creating sample data

The official name of sample data created by a program is **synthetic data**. This subsection is going to present a Python script that creates synthetic data based on given parameters. The logic of the program is based on randomly generated numbers – as most of you might know, randomly generated numbers are not that random. This makes them good for testing the performance of a program but not the actual use of it. However, for the purposes of this book, synthetic data generated with the help of random numbers is fine!

The code of the `synthetic_data.py` Python script is as follows:

```python
#!/usr/bin/env python3

import random
import sys

precision = 5

if len(sys.argv) != 4:
    print("N MIN MAX")
    sys.exit()

# Number of values
N = int(sys.argv[1])
# Minimum value
MIN = int(sys.argv[2])
```

```
# Maximum value
MAX = int(sys.argv[3])

x = random.uniform(MIN, MAX)
# Random float number
for i in range(N):
    print(round(random.uniform(MIN, MAX), precision))
```

The script accepts three parameters, which are the *number of floating-point values* to create the minimum value and the maximum value. Running the script generates the following kind of output:

```
$ ./synthetic_data.py 5 1 3
1.18243
2.81486
1.74816
1.42797
2.21639
```

As floating-point values can have any precision you want, the `precision` variable holds the number of digits after the decimal point that will be printed.

Creating your own time series is not the only way to get data. Publicly available time series data also exists. Let us check this out next.

Publicly available time series data

Websites exist that offer samples of time series data, allowing everyone to process real-world time series data. Another important aspect of publicly available time series data is that people can compare the performance of their techniques, using the same datasets, with others. This is a huge issue in academia, where people have to prove that their techniques and algorithms are faster or more efficient in a variety of aspects compared to others.

A very popular set of publicly available time series data files can be found at https://www.cs.ucr. edu/~eamonn/time_series_data_2018/ (the UCR Time Series Classification Archive).

How time series are processed

Time series processing in Python usually follows these steps:

1. **Importing to Python**: In this step, we import the time series into Python. There exist multiple ways to do that, including reading from a local file, a database server, or an Internet location. In this book, we include all used time series in the GitHub repository as plain text files, which are compressed to save disk space.

2. **Converting it into a time series**: In this step, we convert the data we read in the previous step into a valid time series format. This mainly depends on the Python package used to store the time series data.

3. **Dealing with missing values**: In this step, we look for missing values and ways to deal with them. We are not going to deal with missing values in this book. All presented time series are complete.

4. **Processing time series**: This last step involves processing a time series in order to perform the desired task or tasks.

Reading time series from disk

After storing a time series in a file, we need to write the necessary Python code to read it and put it in a Python variable of some type. This section will teach you exactly that. The read_ts.py script contains the following code:

```python
#!/usr/bin/env python3

import pandas as pd
import numpy as np
import sys

def main():
        filename = sys.argv[1]
        ts1Temp = pd.read_csv(filename, header = None)

        # Convert to NParray
        ta = ts1Temp.to_numpy()
        ta = ta.reshape(len(ta))

        print("Length:", len(ta))

if __name__ == '__main__':
        main()
```

After reading the time series, read_ts.py prints the number of elements in the time series:

```
$ ./read_ts.py ts2
Length: 50
```

The pd.read_csv() function reads a plain text file that uses the CSV format – in our case, each value is on its own line, so there should be no issues with separating values that reside on the same line. The pd.read_csv() function is able to detect issues in the input file. The return value of pd.read_csv() is a *DataFrame* or *TextParser*. In our case, it is a *DataFrame*.

Putting `.astype(np.float64)` at the end of the `pd.read_csv()` statement is going to ensure that all values are read as floating-point values, even when the entire time series contains integer values only. Additionally, `header = None` ensures that the input does not contain a header line with text or data different from the actual data.

The `ts1Temp.to_numpy()` call converts a *DataFrame* into a NumPy array. So, the return value of `ts1Temp.to_numpy()` is a NumPy array. This is needed because we will work with NumPy arrays.

The `ta.reshape(len(ta))` call gives a new shape to an existing NumPy array without changing the data. This is needed for processing the time series data using the correct shape.

As files with time series can be pretty big, it is a good idea to compress them and use them in compressed format. Fortunately, Pandas can read compressed files with the help of a parameter. This is illustrated in the `read_ts_gz.py` script. The statement that does the job is `pd.read_csv(filename, compression='gzip', header = None).astype(np.float64)`. Here, you also see `.astype(np.float64)` in action.

> **How to store time series**
>
> This book uses plain text files to store time series. In these files, each value is in a separate line. More ways exist to store a time series, including the CSV format and the JSON format.

Is all data numeric?

Not all data is numeric, but in time series, almost all data is numeric. The presented script reads a plain text file and makes sure that all data is numeric – note that the `isNumeric.py` script does not currently support compressed files, as it uses the `open()` call to read the input file and it expects a single value per line.

The code of `isNumeric.py` is as follows:

```
#!/usr/bin/env python3

import sys

def main():
    if len(sys.argv) != 2:
        print("TS")
        sys.exit()

    TS = sys.argv[1]
    file = open(TS, 'r')
    Lines = file.readlines()

    count = 0
```

```
    for line in Lines:
        # Strips the newline character
        t = line.strip()
        try:
            _ = float(t)
        except:
            count = count + 1

    print("Number of errors:", count)

if __name__ == '__main__':
    main()
```

The `try` and `except` block is where we try to convert the current string value into a floating-point value using `float()`. If this fails, we know that we are not dealing with a valid numerical value.

Running `isNumeric.py` produces the following kind of output:

```
$ cat ts.txt
5.2
-12.4
-       # Error
17.9
a a     # Error
2 3 4   # Error
4.2
$ ./isNumeric.py ts.txt
Number of errors: 3
```

As we have three erroneous lines, the result is correct.

Do all lines have the same amount of data?

In this subsection, we present a script that counts the number of words in each line, checks that each word is a valid float value, and checks that each line has the same number of values. If not, it states the expected and found number of fields. Additionally, it considers the first line read as the correct one, so all the following lines should have the same amount of data fields. Values are separated by whitespace characters.

The code of `floats_per_line.py` is the following:

```
#!/usr/bin/env python3

import sys

def main():
```

```
        if len(sys.argv) != 2:
            print("TS")
            sys.exit()

        TS = sys.argv[1]
        file = open(TS, 'r')
        Lines = file.readlines()

        first = True
        wordsPerLine = 0
        for line in Lines:
            t = line.strip()
            words = t.split()
            for word in words:
                try:
                    _ = float(word)
                except:
                    print("Error:", word)

            if first:
                wordsPerLine = len(words)
                first = False
            elif wordsPerLine != len(words):
                print("Expected", wordsPerLine, "found", len(words))
                continue

if __name__ == '__main__':
    main()
```

If `String.split()` is executed without any arguments, it splits the string using all the whitespace characters as delimiters, which is what we do here to separate the fields of each input line. If your data is in a different format, you might need to modify the `String.split()` statement to match your needs.

Running `floats_per_line.py` produces the following kind of output:

```
$ ./floats_per_line.py ts.txt
Error: -
Error: a
Error: b
Expected 1 found 2
Expected 1 found 3
```

The next subsection shows how to process time series based on the sliding window size.

Creating subsequences

Although we read a time series as an entity from a plain text file, we process it as a large list of subsequences. In this subsection, you are going to learn how to process a time series as a list of subsequences, based on a given sliding window size.

The name of the Python script is subsequences.py. We are going to present it in two parts. Among other things, the first part contains the Python structure used to keep the subsequences:

```python
#!/usr/bin/env python3

import argparse
import stumpy
import numpy as np
import pandas as pd
import sys

class TS:
    def __init__(self, ts, index):
        self.ts = ts
        self.index = index
```

The TS class has two members, one to hold the actual data (the ts variable) and the other to keep the index (index variable) number of the subsequence. The chapters that follow are going to enrich the TS class to fit our growing needs.

The rest of the script is the following:

```python
def main():
    parser = argparse.ArgumentParser()
    parser.add_argument("-w", dest = "window", type=int)
    parser.add_argument("TS")

    args = parser.parse_args()
    windowSize = args.window
    file = args.TS

    ts = pd.read_csv(file, names=['values'], compression='gzip',
header = None)
    ts_numpy = ts.to_numpy()
    length = len(ts_numpy)

    # Split time series into subsequences
    for i in range(length - windowSize + 1):
        # Get the subsequence
        ts = ts_numpy[i:i+windowSize]
```

```
        # Create new TS node based on ts
        ts_node = TS(ts, i)


if __name__ == '__main__':
    main()
```

The `argparse` package helps us put the command-line arguments in order. In this case, we expect two parameters: first, the sliding window size (`-w`), and second, the filename that holds the time series. The `for` loop is used to split the time series into subsequences and generate multiple TS class members.

The previous code is not something difficult to read or understand or make changes to. Expect to see this sort of code in most of the Python scripts in this book!

In its current form, `subsequences.py` generates no output. You will only get error messages if something is wrong with the provided filename or its data.

Visualizing time series

Most of the time, having a high-level overview of your data is an excellent way to get to know your data. The best way to get an overview of a time series is by visualizing it.

There are multiple ways to visualize a time series, including tools such as R or Matlab, or using a large amount of existing JavaScript packages. In this section, we are going to use a Python package called Matplotlib for visualizing the data. Additionally, we will save the output to a PNG file. A viable alternative to this is to use a Jupyter notebook – Jupyter comes with Anaconda – and display the graphical output on your favorite web browser.

The `visualize.py` script reads a plain text file with values – a time series – and creates a plot. The Python code of `visualize.py` is as follows:

```
#!/usr/bin/env python3

import sys
import pandas as pd
import matplotlib.pyplot as plt
import numpy as np
import math

def main():
    if len(sys.argv) != 2:
        print("TS")
        sys.exit()

    F = sys.argv[1]
```

```
    # Read Sequence as Pandas
    ts = pd.read_csv(F, compression='gzip', header = None)
    # Convert to NParray
    ta = ts.to_numpy()
    ta = ta.reshape(len(ta))

    plt.plot(ta, label=F, linestyle='-', markevery=100, marker='o')
    plt.xlabel('Time Series', fontsize=14)
    plt.ylabel('Values', fontsize=14)

    plt.grid()
    plt.savefig("CH01_03.png", dpi=300, format='png', bbox_
inches='tight')

if __name__ == '__main__':
    main()
```

You must be familiar with most of the presented code, as you already saw some of it earlier in this chapter. The `plt.plot()` statement plots the data, whereas the `plt.savefig()` function saves the output in the file instead of displaying it on screen.

The output from the `./visualize.py ts1.gz` command can be seen in *Figure 1.3*:

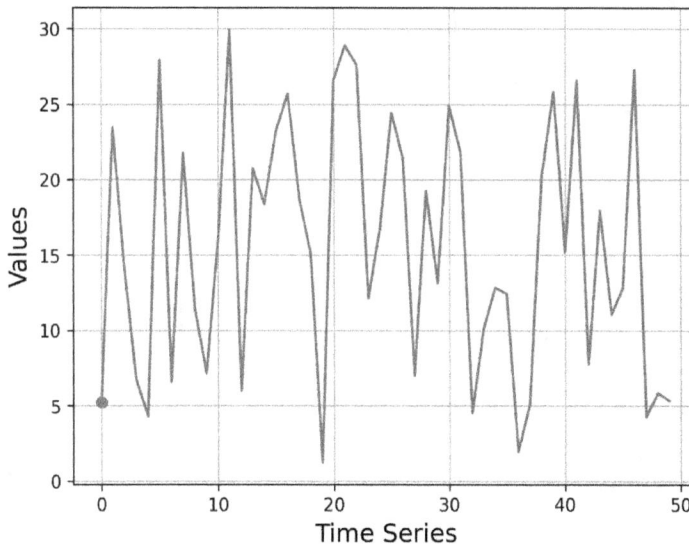

Figure 1.3 – Visualizing a time series

Now that we understand how to work with time series and subsequences, it is time to present an advanced technique called the **Matrix Profile**, which shows the tasks that we might need to compute when working with time series and how time-consuming these tasks might be.

Working with the Matrix Profile

In this section, as well as the next one, we will work with the `stumpy` Python package. This package is not related to iSAX but offers lots of advanced functionality related to time series. With the help of `stumpy`, we can calculate the *Matrix Profile*.

The Matrix Profile is two things:

- A vector of distances that shows the distance of each subsequence in a time series to its nearest neighbor
- A vector of indexes that shows the index of the nearest neighbor of each subsequence in a time series

The Matrix Profile can be used in many time series mining tasks. The main reason for presenting it is to understand that working with time series can be slow, so we need structures and techniques to improve the performance of time series-related tasks.

To get a better idea of the use of the Matrix Profile and the time it takes `stumpy` to calculate the Matrix Profile, here is the Python code of `matrix_profile.py`:

```python
#!/usr/bin/env python3

import pandas as pd
import argparse

import time

import stumpy

def main():
    parser = argparse.ArgumentParser()
    parser.add_argument("-w", "--window", dest = "window", default =
"16", help="Sliding Window", type=int)
    parser.add_argument("TS")
    args = parser.parse_args()

    windowSize = args.window
    inputTS = args.TS

    print("TS:", inputTS, "Sliding Window size:", windowSize)
```

```
        start_time = time.time()
        ts = pd.read_csv(inputTS, names=['values'], compression='gzip')

        # Convert to NParray
        ts_numpy = ts.to_numpy()
        ta = ts_numpy.reshape(len(ts_numpy))
        realMP = stumpy.stump(ta, windowSize)
        print("--- %.5f seconds ---" % (time.time() - start_time))

if __name__ == '__main__':
    main()
```

The `stumpy.stump()` function calculates the Matrix Profile of a given time series.

We are going to execute `matrix_profile.py` two times. The first time using a time series with 100,000 elements and the second time using a time series with 300,000 elements. As with almost all the Python scripts in this book that read a time series, `matrix_profile.py` expects to read compressed plain text files.

The taskset(1) command

The `taskset(1)` command is used to assign the desired number of cores to a given process and is currently available on Linux machines. The reason for using it is to limit the number of available cores when executing `matrix_profile.py` and `mpdistance.py`, which, by default, use all available cores due to the use of the Numba Python package. As a rule of thumb, when testing the performance of an algorithm or comparing one to another, it is better to use a single core. There is no similar utility on macOS.

Running `matrix_profile.py` with the smaller time series produces the following output:

```
$ taskset --cpu-list 3 ./matrix_profile.py 100k.txt.gz
TS: 100k.txt.gz Sliding Window size: 16
--- 120.44 seconds ---
```

So, it took `stumpy.stump()` around `120.44` seconds to process a time series with 100,000 elements.

Running `matrix_profile.py` with the bigger time series produces the following output:

```
$ taskset --cpu-list 0 ./matrix_profile.py 300k.gz -w 1024
TS: 300k.gz Sliding Window size: 1024
--- 922.30060 seconds ---
```

Here, it took `stumpy.stump()` around `922` `seconds` to process a time series with 300,000 elements on a single CPU core. Now, imagine doing the same with a time series that has more than 1,000,000 elements!

You are going to learn all about the Matrix Profile and understand why it is so slow in *Chapter 7*.

The next section discusses a distance function named **MPdist** that internally uses the Matrix Profile for its computation.

Exploring the MPdist distance

MPdist offers a way to calculate the distance between two time series. Strictly speaking, the *MPdist* distance is a distance measure that is based on the Matrix Profile. It is much slower to compute than the Euclidean distance, but it does not require the time series to have the same size.

As you might expect, it must offer many advantages when compared to the Euclidean distance, as well as other existing distance metrics. The main advantages of MPdist, according to the people that created it, are the following:

- It is more flexible regarding the way it compares data than most existing distance functions.
- It considers similarities of data that may not take place at the same time, where time means at the same index.
- MPdist is considered more robust in specific analytics scenarios due to the way it is computed. More specifically, MPdist is more robust to spikes and missing values.

As MPdist is based on the Matrix Profile, calculating the MPdist distance can be extremely slow, especially when working with large time series.

First, let us have a look at the Python code of mpdistance.py:

```python
#!/usr/bin/env python3

import stumpy
import stumpy.mpdist
import numpy as np

import time
import sys
import pandas as pd

if len(sys.argv) != 4:
    print("TS1 + TS2 + Window size")
    sys.exit()

# Time series files
TS1 = sys.argv[1]
TS2 = sys.argv[2]
windowSize = int(sys.argv[3])
```

```
print("TS1:", TS1, "TS2:", TS2, "Window Size:", windowSize)

# Read Sequence as Pandas
ts1Temp = pd.read_csv(TS1, compression='gzip', header = None).
astype(np.float64)
# Convert to NParray
ta = ts1Temp.to_numpy()
ta = ta.reshape(len(ta))

# Read Sequence as Pandas
ts2Temp = pd.read_csv(TS2, compression='gzip', header = None).
astype(np.float64)
# Convert to NParray
tb = ts2Temp.to_numpy()
tb = tb.reshape(len(tb))

print(len(ta), len(tb))

start_time = time.time()
mpdist = stumpy.mpdist(ta, tb, m=windowSize)
print("--- %.5f seconds ---" % (time.time() - start_time))
print("MPdist: %.4f " % mpdist)
```

This program uses command-line arguments with the help of sys to read the required data, instead of the argparse package.

All of this is done by the call to stumpy.mpdist(), which requires three parameters – the two time series and the sliding window size.

As mpdistance.py calculates the distance between two time series, it expects to read two files. Running mpdistance.py with two synthetic data sets with 100,000 elements each generates the following output:

```
$ taskset --cpu-list 0 ./mpdistance.py 100k_1.txt.gz 100k_2.txt.gz 512
TS1: 100k_1.txt.gz TS2: 100k_2.txt.gz Window Size: 512
100000 100000
--- 349.81955 seconds ---
MPdist: 28.3882
```

So, on a Linux machine with an Intel i7 CPU, it took mpdistance.py 349.81955 seconds to execute when *using a single CPU core*. The value of the MPdist distance is 28.3882.

If we use two datasets with half a million elements each (500,000 elements), the output and the time it takes to execute `mpdistance.py` should be similar to the following:

```
$ taskset --cpu-list 3 ./mpdistance.py h_500k_f.gz t_500k_f.gz 2048
TS1: h_500k_f.gz TS2: t_500k_f.gz Window Size: 2048
506218 506218
--- 4102.92 seconds ---
MPdist: 38.2851
```

So, on a Linux machine with an Intel i7 CPU, it took `mpdistance.py` 4102.92 seconds to execute when *using a single CPU core*. The value of the MPdist distance is 38.2851.

You will learn all the details about MPdist in *Chapter 7*. For now, what you should keep in mind is that MPdist is a distance function that has some performance issues.

> **Experiment and stay humble**
>
> If I can give you just one piece of advice to remember from this book, it would be to experiment and try things. Experiment as much as possible with what you read, question it, think in new ways, try new things, and keep learning. At the same time, stay humble, and do not forget that lots of people have worked to lay the foundation for us to be here today talking about time series and indexing.

Summary

In this chapter, we learned the basics of time series, indexing, and distance functions. Although the theoretical knowledge included in this chapter remains valid and relevant no matter what programming language is used, alternative ways and packages to achieve the presented tasks in Python exist, as well as other programming languages. What remains unchanged is the validity of the approach – you must read a text file from disk in order to use its data, no matter the programming language – and the logical steps needed to perform a task, such as the visualization of a time series. This means that if you know an alternative way to load a text file from disk in Python, feel free to use it if it allows you to perform the next task at hand. If you are an amateur Python developer, I would suggest that you follow the book's suggestions until you become more competent with Python. After all, the used Python packages are the most popular ones in the Python community.

Before you continue reading this book, please make sure that you understand the knowledge presented in this chapter, as it is the foundation for the rest of the book, especially if you are new to time series and indexing.

The next chapter is about the SAX representation, which is an integral part of the iSAX index.

Resources and useful links

- The Stumpy Python package: `https://pypi.org/project/stumpy/`.

- The NumPy Python package: `https://numpy.org/`.

- The SciPy Python package: `https://pypi.org/project/scipy/` and `https://scipy.org/`.

- Anaconda documentation: `https://docs.anaconda.com/`.

- Anaconda distribution: `https://www.anaconda.com/products/distribution`.

- Jupyter Notebooks: `https://jupyter.org/`.

- Matplotlib: `https://pypi.org/project/matplotlib/` and `https://matplotlib.org/`.

- The Z Shell: `https://www.zsh.org/`.

- Oh My Zsh: `https://github.com/ohmyzsh/ohmyzsh`.

- The Pandas Python package: `https://pypi.org/project/pandas/`.

- The Dewey Decimal Classification system: `https://en.wikipedia.org/wiki/Dewey_Decimal_Classification`.

- The Numba Python package: `https://pypi.org/project/numba/`.

- The Matrix Profile is defined in a research paper called *Matrix Profile I: All Pairs Similarity Joins for Time Series: A Unifying View That Includes Motifs, Discords and Shapelets*. The authors of the paper are Chin-Chia Michael Yeh, Yan Zhu, Liudmila Ulanova, Nurjahan Begum, Yifei Ding, Hoang Anh Dau, Diego Furtado Silva, Abdullah Mueen, and Eamonn J. Keogh.

- The MPdist distance is defined in a research paper called *Matrix Profile XII: MPdist: A novel time series distance measure to allow data mining in more challenging scenarios*. The authors of the paper are S. Gharghabi, S. Imani, A. Bagnall, A. Darvishzadeh, and E. Keogh.

- You can find more information about `numpy.linalg.norm()` at `https://numpy.org/doc/stable/reference/generated/numpy.linalg.norm.html`.

- The Homebrew macOS package manager: `https://brew.sh/`.

Exercises

Try to do the following exercises:

- Create a new Anaconda environment.

- List the installed packages of an Anaconda environment.

- Delete an existing Anaconda environment.

- Create a new synthetic dataset with 1,000 values from -10 to +10.

- Create a new synthetic dataset with 100,000 values from 0 to +10.

- Write a Python script that reads a plain text file line by line.

- Write a Python script that reads a plain text file and prints it word by word. Why is this more difficult than printing a file line by line?

- Write a Python script that reads the same plain text file multiple times, and time that operation. The number of times the file is read as well as the file path should be given as command-line arguments.

- Modify `synthetic_data.py` to generate integer values instead of floating-point values.

- Create a time series with 500,000 elements with `synthetic_data.py`, and execute `matrix_profile.py` on the generated time series. Do not forget to compress the plain text file.

- Modify `mpdistance.py` to use `argparse` to read its parameters.

- Experiment with `visualize.py` to plot your own time series. What happens when you plot big time series? How easy it is to find what you are looking for?

- In the iSAX index in *Figure 1.2* a binary tree? Is it a balanced tree? Why?

- Modify `ed.py` to read the time series from compressed plain text files.

2
Implementing SAX

This chapter is about the **Symbolic Aggregate Approximation** (**SAX**) component of the iSAX index and is divided into two parts – the first part with the theoretical knowledge, and the second part with the code to compute SAX and the practical examples. At the end of the chapter, you will see how to calculate some handy statistical quantities that can give you a higher overview of your time series and plot a histogram of your data.

In this chapter, we will cover the following main topics:

- The required theory
- An introduction to SAX
- Developing a Python package
- Working with the SAX package
- Counting the SAX representations of a time series
- The `tsfresh` Python package
- Creating a histogram of a time series
- Calculating the percentiles of a time series

Technical requirements

The GitHub repository for the book is `https://github.com/PacktPublishing/Time-Series-Indexing`. The code for each chapter is in its own directory. Therefore, the code for *Chapter 2* can be found in the `ch02` folder. If you already used `git(1)` to get a local copy of the entire GitHub repository, there is no need to get that again. Just make your current working directory `ch02` while working with this chapter.

The required theory

In this section, you are going to learn the required theory that supports the SAX representation. However, keep in mind that this book is more practical than it is theoretical. If you want to learn the theory in depth, you should read the research papers mentioned in this chapter, as well as the forthcoming ones, and the *Useful links* section found at the end of each chapter. Thus, the theory is about serving our main purpose, which is the implementation of techniques and algorithms.

The operation and the details of SAX are fully described in a research paper titled *Experiencing SAX: a novel symbolic representation of time series*, which was written by Jessica Lin, Eamonn Keogh, Li Wei, and Stefano Lonardi. This paper (`https://doi.org/10.1007/s10618-007-0064-z`) was officially published back in 2007. You do not have to read all of it from the front cover to the back cover, but it is a great idea to download it and read the first pages of it, giving special attention to the abstract and the introduction section.

We will begin by explaining the terms *PAA* and *SAX*. **PAA** stands for **Piecewise Aggregate Approximation**. The PAA representation offers a way to reduce the dimensionality of a time series. This means that it takes a long time series and creates a smaller version of it that is easier to work with.

PAA is also explained in the *Experiencing SAX: a novel symbolic representation of time series* paper (`https://doi.org/10.1007/s10618-007-0064-z`). From that, we can easily understand that PAA and SAX are closely related, as the idea behind SAX is based on PAA. The *SAX representation* is a **symbolic representation of time series**. Put simply, it offers a way of representing a time series in a summary form, in order to save space and increase speed.

> **The difference between PAA and SAX**
>
> The main difference between PAA and the SAX representation is that PAA just calculates the mean values of a time series, based on a sliding window size, whereas the SAX representation utilizes those mean values and further transforms PAA to get a discrete representation of a time series (or subsequence). In other words, the SAX representation converts the PAA representation into something that is better to work with. As you will find out in a while, this transformation takes place with the help of **breakpoints**, which divide the numeric space of the mean values into subspaces. Each subspace has a discrete representation based on the given breakpoint values.

Both PAA and SAX are techniques for dimensionality reduction. SAX is going to be explained in much more detail in a while, whereas the discussion about PAA ends here.

The next subsection tells us why we need SAX.

Why do we need SAX?

Time series are difficult to search. The longer a time series (or subsequence) is, the more computationally intensive it is to search for it or compare it with another one. The same applies to working with indexes that index time series – iSAX is such an index.

To make things simpler for you, what we will do is take a subsequence with x elements and transform it into a representation with w elements, where w is much smaller than x. In strict terms, this is called **dimensionality reduction**, and it allows us to work with long subsequences using less data. However, once we decide that we need to work with a given subsequence, we need to work with it using its full dimensions – that is, all its x elements.

The next subsection talks about normalization, which, among other things, allows us to compare values at different scales.

Normalization

The first two questions you might ask are what normalization is and why we need it.

Normalization is the process of adjusting values that use different scales to a common scale. A simple example is comparing Fahrenheit and Celsius temperatures – we cannot do that unless we bring all values to the same scale. This is the simplest form of normalization.

Although various types of normalization exist, what is needed here is **standard score normalization**, which is the simplest form of normalization, because this is what is used for time series and subsequences. Please do not confuse database normalization and normal forms with value normalization, as they are totally different concepts.

The reasons that we introduce normalization into the process are as follows:

- The first and most important reason is that we can compare datasets that use a different range of values. A simple case is comparing Celsius and Fahrenheit temperatures.
- A side effect of the previous point is that data anomalies are reduced but not eliminated.
- In general, normalized data is easier to understand and process because we deal with values in a predefined range.
- Searching using an index that uses normalized values might be faster than when working with bigger values.
- Searching, sorting, and creating indexes is faster since values are smaller.
- Normalization is conceptually cleaner and easier to maintain and change as your needs change.

Another simple example that supports the need for normalization is when comparing positive values with negative ones. It is almost impossible to draw useful conclusions when comparing such different kinds of observations. Normalization solves such issues.

Although we are not going to need to, bear in mind that we cannot go from the normalized version of a subsequence to the original subsequence, so the normalization process is irreversible.

The following function shows how to normalize a time series with some help from the NumPy Python package:

```
def normalize(x):
    eps = 1e-6
    mu = np.mean(x)
    std = np.std(x)
    if std < eps:
            return np.zeros(shape=x.shape)
    else:
            return (x-mu)/std
```

The previous function reveals the formula of normalization. Given a dataset, the normalized form of each one of its elements is equal to the value of the observation, minus the **mean value** of the dataset over the **standard deviation** of the dataset – both these statistical terms are explained in *The tsfresh Python package* section of this chapter.

This is seen in the return value of the previous function, `(x-mu)/std`. NumPy is clever enough to calculate that value for each observation without the need to use a `for` loop. If the standard deviation is close to 0, which is simulated by the value of the `eps` variable, then the return value of `normalize()` is equal to a NumPy array full of zeros.

The `normalize.py` script, which uses the previously developed function that does not appear here, gets a time series as input and returns its normalized version. Its code is as follows:

```
#!/usr/bin/env python3

import sys
import pandas as pd
import numpy as np

def main():
    if len(sys.argv) != 2:
            print("TS")
            sys.exit()

    F = sys.argv[1]

    ts = pd.read_csv(F, compression='gzip', header = None)
    ta = ts.to_numpy()
    ta = ta.reshape(len(ta))
```

```
        taNorm = normalize(ta)

        print("[", end = ' ')
        for i in taNorm.tolist():
                print("%.4f" % i, end = ' ')
        print("]")

if __name__ == '__main__':
        main()
```

The last `for` loop of the program is used to print the contents of the `taNorm` NumPy array with a smaller precision in order to take up less space. To do that, we need to convert the `taNorm` NumPy array into a regular Python list using the `tolist()` method.

We are going to feed `normalize.py` a short time series; however, the script also works with longer ones. The output of `normalize.py` looks as follows:

```
$ ./normalize.py ts1.gz
[ -1.2272 0.9487 -0.1615 -1.0444 -1.3362 1.4861 -1.0620 0.7451 -0.4858
-0.9965 0.0418 1.7273 -1.1343 0.6263 0.3455 0.9238 1.2197 0.3875
-0.0483 -1.7054 1.3272 1.5999 1.4479 -0.4033 0.1525 1.0673 0.7019
-1.0114 0.4473 -0.2815 1.1239 0.7516 -1.3102 -0.6428 -0.3186 -0.3670
-1.6163 -1.2383 0.5692 1.2341 -0.0372 1.3250 -0.9227 0.2945 -0.5290
-0.3187 1.4103 -1.3385 -1.1540 -1.2135 ]
```

With normalization in mind, let us now proceed to the next subsection, where we are going to visualize a time series and show the visual difference between the original version and the normalized version of it.

Visualizing normalized time series

In this subsection, we are going to show the difference between the normalized and the original version of a time series with the help of visualization. Keep in mind that we usually *do not normalize the entire time series*. The normalization takes place on a subsequence level based on the sliding window size. In other words, for the purposes of this book, we will normalize subsequences, not an entire time series. Additionally, for the calculation of the SAX representation, we process the normalized subsequences based on the **segment value**, which specifies the parts that a SAX representation will have. So, for a segment value of 2, we split the normalized subsequence into two. For a segment value of 4, we split the normalized subsequence into four sets.

Nevertheless, viewing the normalized and original versions of a time series is very educational. The Python code of `visualize_normalized.py`, without the implementation of `normalize()`, is as follows:

```python
#!/usr/bin/env python3

import sys
import pandas as pd
import matplotlib.pyplot as plt
import numpy as np

def main():
    if len(sys.argv) != 2:
        print("TS")
        sys.exit()

    F = sys.argv[1]

    ts = pd.read_csv(F, compression='gzip', header = None)
    ta = ts.to_numpy()
    ta = ta.reshape(len(ta))

    # Find its normalized version
    taNorm = normalize(ta)

    plt.plot(ta, label="Regular", linestyle='-', markevery=10,
marker='o')
    plt.plot(taNorm, label="Normalized", linestyle='-.',
markevery=10, marker='o')
    plt.xlabel('Time Series', fontsize=14)
    plt.ylabel('Values', fontsize=14)
    plt.legend()

    plt.grid()
    plt.savefig("CH02_01.png", dpi=300, format='png', bbox_
inches='tight')

if __name__ == '__main__':
    main()
```

The `plt.plot()` function is called twice, plotting a line each time. Feel free to experiment with the Python code in order to change the look of the output.

Figure 2.1 shows the output of `visualize_normalized.py ts1.gz`, which uses a time series with 50 elements.

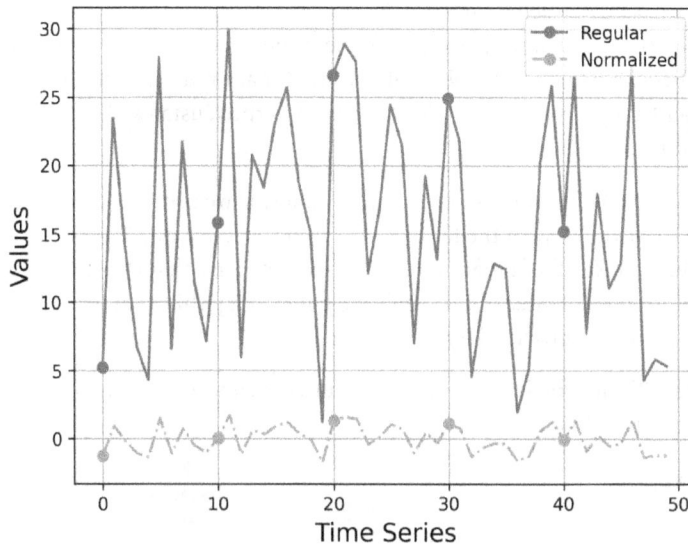

Figure 2.1 – The plotting of a time series and its normalized version

I think that *Figure 2.1* speaks for itself! The values of the normalized version are located *around the value of 0*, whereas the values of the original time series can be anywhere! Additionally, we make the original time series smoother without completely losing its original shape and edges.

The next section is about the details of the SAX representation, which is a key component of every iSAX index.

An introduction to SAX

As mentioned previously, **SAX** stands for **Symbolic Aggregate Approximation**. The SAX representation was officially announced back in 2007 in the *Experiencing SAX: a novel symbolic representation of time series* paper (`https://doi.org/10.1007/s10618-007-0064-z`).

Keep in mind that we do not want to find the SAX representation of an entire time series. We just want to find the SAX representation of a subsequence of a time series. The main difference between a time series and a subsequence is that a time series is many times bigger than a subsequence.

Each SAX representation has two parameters named **cardinality** and the number of **segments**. We will begin by explaining the cardinality parameter.

The cardinality parameter

The *cardinality* parameter specifies the number of possible values each segment can have. As a side effect, the cardinality parameter *defines the way the y axis is divided* – this is used to get the value of each segment. There exist multiple ways to specify the value of a segment based on the cardinality. These include alphabet characters, decimal numbers, and binary numbers. In this book, we will use binary numbers because they are easier to understand and interpret, using a file with the **precalculated breakpoints** for cardinalities up to 256.

So, a cardinality of 4, which is 2^2, gives us four possible values, as we use 2 bits. However, we can easily replace 00 with the letter a, 01 with the letter b, 10 with the letter c, 11 with the letter d, and so on, in order to use letters instead of binary numbers. Keep in mind that this might require minimal code changes in the presented code, and it would be good to try this as an exercise when you feel comfortable with SAX and the provided Python code.

The format of the file with the breakpoints, which in our case supports cardinalities up to 256 and is called SAXalphabet, is as follows:

```
$ head -7 SAXalphabet
0
-0.43073,0.43073
-0.67449,0,0.67449
-0.84162,-0.25335,0.25335,0.84162
-0.96742,-0.43073,0,0.43073,0.96742
-1.0676,-0.56595,-0.18001,0.18001,0.56595,1.0676
-1.1503,-0.67449,-0.31864,0,0.31864,0.67449,1.1503
```

The values presented here are called breakpoints in the SAX terminology. The value in the first line divides the *y* axis into two areas, separated by the *x* axis. So, in this case, we need 1 bit to define whether we are in the upper space (the positive *y* value) or the lower one (the negative *y* value).

As we will use binary numbers to represent each SAX segment, there is no point in wasting them. Therefore, the values that we will use are the powers of 2, from 2^1 (*cardinality 2*) to 2^8 (*cardinality 256*).

Let us now present *Figure 2.2*, which shows how -0.67449, 0, 0.67449 divides the y axis, which is used in the 2^2 cardinality. The bottom part begins from the minus infinitive up to -0.67449, the second part from -0.67449 up to 0, the third part from 0 to 0.67449, and the last part from 0.67449 up to plus infinitive.

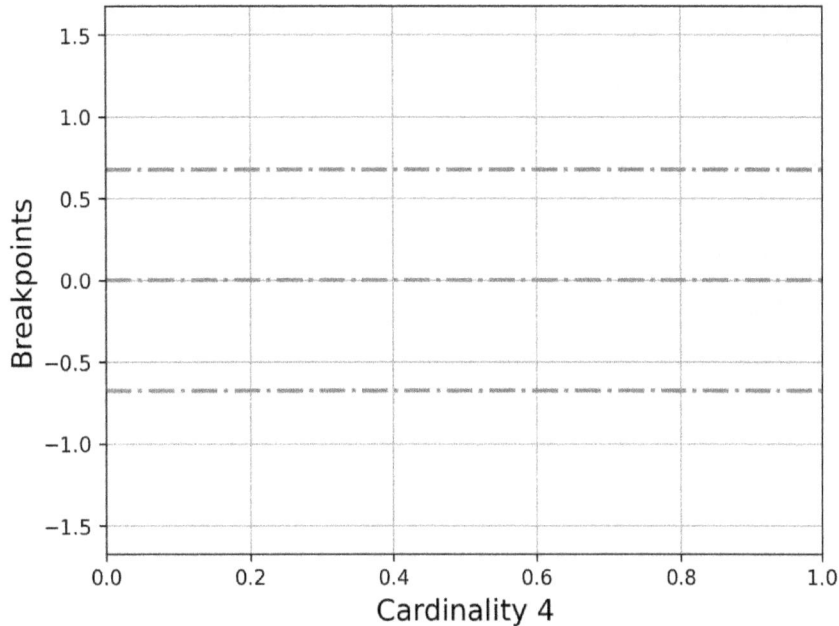

Figure 2.2 – The y axis for cardinality 4 (three breakpoints)

Let us now present *Figure 2.3*, which shows how -1.1503, -0.67449, -0.31864, 0, 0.31864, 0.67449, 1.1503 divides the y axis. This is for the 2^3 cardinality.

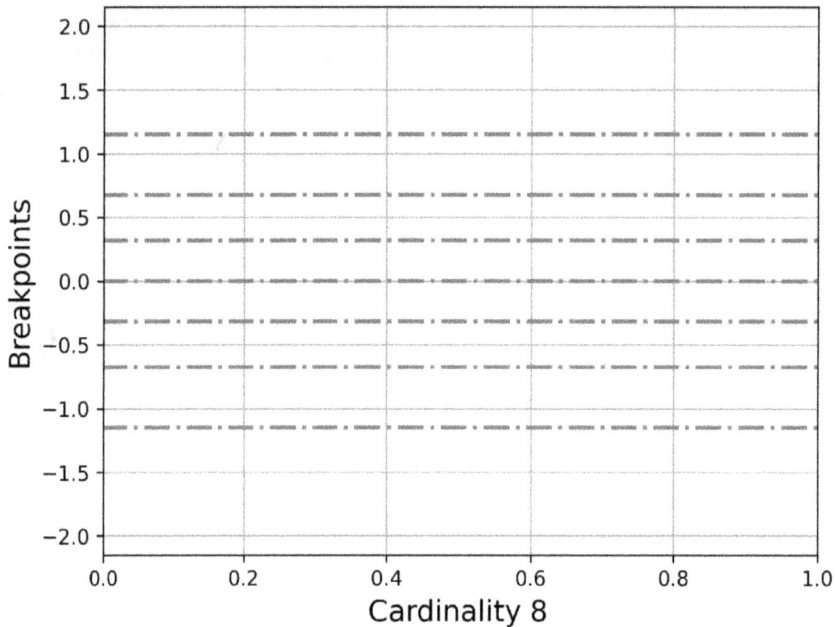

Figure 2.3 – The y axis for cardinality 8 (7 breakpoints)

As this can be a tedious job, we have created a utility that does all the plotting. Its name is `cardinality.py`, and it reads the `SAXalphabet` file to find the breakpoints of the desired cardinality before plotting them.

The Python code for `cardinality.py` is as follows:

```python
#!/usr/bin/env python3

import sys
import pandas as pd
import matplotlib.pyplot as plt
import numpy as np
import os

breakpointsFile = "./sax/SAXalphabet"

def main():
    if len(sys.argv) != 3:
        print("cardinality output")
        sys.exit()
```

```
n = int(sys.argv[1]) - 1
output = sys.argv[2]

path = os.path.dirname(__file__)
file_variable = open(path + "/" + breakpointsFile)
alphabet = file_variable.readlines()
myLine = alphabet[n - 1].rstrip()
elements = myLine.split(',')

lines = [eval(i) for i in elements]

minValue = min(lines) - 1
maxValue = max(lines) + 1

fig, ax = plt.subplots()

for i in lines:
        plt.axhline(y=i, color='r', linestyle='-.', linewidth=2)

xLabel = "Cardinality " + str(n)

ax.set_ylim(minValue, maxValue)
ax.set_xlabel(xLabel, fontsize=14)
ax.set_ylabel('Breakpoints', fontsize=14)

ax.grid()
fig.savefig(output, dpi=300, format='png', bbox_inches='tight')

if __name__ == '__main__':
    main()
```

The script requires two command-line parameters – the cardinality and the output file, which is used to save the image. Note that a cardinality value of 8 requires 7 breakpoints, a cardinality value of 32 requires 31 breakpoints, and so on. Therefore, the Python code for cardinality.py decreases the line number that it is going to search for in the SAXalphabet file to support that functionality. Therefore, when given a cardinality value of 8, the script is going to look for the line with 7 breakpoints in SAXalphabet. Additionally, as the script reads the breakpoint values as strings, we need to convert these strings into floating-point values using the lines = [eval(i) for i in elements] statement. The rest of the code is related to the Matplotlib Python package and how to draw lines using plt.axhline().

The next subsection is about the segments parameter.

The segments parameter

The (number of) *segments* parameter specifies the number of parts (*words*) a SAX representation is going to have. Therefore, a segments value of 2 means that the SAX representation is going to have two words, each one using the specified cardinality. Therefore, the values of each part are determined by the cardinality.

A side effect of this parameter is that, after normalizing a subsequence, we divide it by the number of segments and work with these different parts separately. This is the way the SAX representation works.

Both cardinality and segments values control the data compression ratio and the accuracy of the subsequences of a time series and, therefore, the entire time series.

The next subsection shows how to manually compute the SAX representation of a subsequence – this is the best way to fully understand the process and be able to identify bugs or errors in the code.

How to manually find the SAX representation of a subsequence

Finding the SAX representation of a subsequence looks easy but requires lots of computations, which makes the process ideal for a computer. Here are the steps to find the SAX representation of a time series or subsequence:

1. First, we need to have the number of segments and the cardinality.
2. Then, we normalize the subsequence or the time series.
3. After that, we divide the normalized subsequence by the number of segments.
4. For each one of these parts, we find its mean value.
5. Finally, based on each mean value, we calculate its representation based on the cardinality. The cardinality is what defines the breakpoint values that are going to be used.

We will use two simple examples to illustrate the manual computation of the SAX representation of a time series. The time series is the same in both cases. What will be different are the SAX parameters and the sliding window size.

Let's imagine we have the following time series and a sliding window size of 4:

```
{-1, 2, 3, 4, 5, -1, -3, 4, 10, 11, . . .}
```

Based on the sliding window size, we extract the first two subsequences from the time series:

- S1 = {-1, 2, 3, 4}
- S2 = {2, 3, 4, 5}

The first step that we should take is to **normalize these two subsequences**. For that, we will use the `normalize.py` script we developed earlier – we just have to save each subsequence into its own plain text file and compress it using the `gzip` utility, before giving it as input to `normalize.py`. If you use a Microsoft Windows machine, you should look for a utility that allows you to create such ZIP files. An alternative is to work with plain text files, which might require some small code changes in the `pd.read_csv()` function call.

The output of the `normalize.py` script when processing S1 (`s1.txt.gz`) and S2 (`s2.txt.gz`) is the following:

```
$ ./normalize.py s1.txt.gz
[ -1.6036 0.0000 0.5345 1.0690 ]
$ ./normalize.py s2.txt.gz
[ -1.3416 -0.4472 0.4472 1.3416 ]
```

So, the normalized versions of S1 and S2 are as follows:

- N1 = {-1.6036, 0.0000, 0.5345, 1.0690}
- N2 = {-1.3416, -0.4472, 0.4472, 1.3416}

In this first example, we use a segments value of 2 and a cardinality value of 4 (22). A segment value of 2 means that we must divide each *normalized subsequence* into two parts. These two parts contain the following data, based on the normalized versions of S1 and S2:

- For S1, the two parts are {-1.6036, 0.0000} and {0.5345, 1.0690}
- For S2, the two parts are {-1.3416, -0.4472} and {0.4472, 1.3416}

The mean values of each part are as follows:

- For S1, they are -0.8018 and 0.80175
- For S2, they are -0.8944 and 0.8944

For the cardinality of 4, we are going to look at *Figure 2.2* and the respective breakpoints, which are -0.67449, 0, and 0.67449. So, the SAX values of each segment are as follows:

- For S1, they are 00 because -0.8018 falls at the bottom of the plot and 11
- For S2, they are 00 and 11 because 0.8944 falls at the top of the plot

Therefore, the SAX representation of S1 is [00, 11] and for S2, it is [00, 11]. It turns out that both subsequences have the same SAX representation. This makes sense, as they only differ in one element, which means that their normalized versions are similar.

Note that in both cases, the lower cardinality begins from the bottom of the plot. For *Figure 2.2*, this means that 00 is at the bottom of the plot, 01 is next, followed by 10, and 11 is at the top of the plot.

In the second example, we will use a sliding window size of 8, a segments value of 4, and a cardinality value of 8 (23).

About the sliding window size

Keep in mind that the normalized representation of the subsequence remains the same when the sliding window size remains the same. However, if either the cardinality or the segments change, the resulting SAX representation might be completely different.

Based on the sliding window size, we extract the first two subsequences from the time series – S1 = {-1, 2, 3, 4, 5, -1, -3, 4} and S2 = {2, 3, 4, 5, -1, -3, 4, 10}.

The output of the normalize.py script is going to be the following:

```
$ ./normalize.py T1.txt.gz
[ -0.9595 0.1371 0.5026 0.8681 1.2337 -0.9595 -1.6906 0.8681 ]
$ ./normalize.py T2.txt.gz
[ -0.2722 0.0000 0.2722 0.5443 -1.0887 -1.6330 0.2722 1.9052 ]
```

So, the normalized versions of S1 and S2 are N1 = {-0.9595, 0.1371, 0.5026, 0.8681, 1.2337, -0.9595, -1.6906, 0.8681} and N2 = {-0.2722, 0.0000, 0.2722, 0.5443, -1.0887, -1.6330, 0.2722, 1.9052}, respectively.

A segment value of 4 means that we must divide each one of the *normalized subsequences* into four parts. For S1, these parts are {-0.9595, 0.1371}, {0.5026, 0.8681}, {1.2337, -0.9595}, and {-1.6906, 0.8681}.

For S2, these parts are {-0.2722, 0.0000}, {0.2722, 0.5443}, {-1.0887, -1.6330}, and {0.2722, 1.9052}.

For S1, the mean values are -0.4112, 0.68535, 0.1371, and -0.41125. For S2, the mean values are -0.1361, 0.40825, -1.36085, and 1.0887.

About the breakpoints of the cardinality value of 8

Just a reminder here that for the cardinality value of 8, the breakpoints are (000) -1.1503, (001) -0.67449, (010) -0.31864, (011) 0, (100) 0.31864, (101) 0.67449, and (110) 1.1503 (111). In parentheses, we present the SAX values for each breakpoint. For the first breakpoint, we have the 000 value to its left and 001 to its right. For the last breakpoint, we have the 110 value to its left and 111 to its right. Remember that we use seven breakpoints for a cardinality value of 8.

Therefore, the SAX representation of S1 is ['010', '110', '100', '010'], and for S2, it is ['011', '101', '000', '110']. The use of single quotes around SAX words means that internally we store SAX words as strings, despite the fact that we calculate them as binary numbers because it is easier to search and compare strings.

The next subsection examines a case where a subsequence cannot be divided perfectly by the number of segments.

How can we divide 10 data points into 3 segments?

So far, we have seen examples where the length of the subsequence can be perfectly divided by the number of segments. However, what happens if that is not possible?

In that case, there exist data points that contribute to two adjacent segments at the same time. However, instead of putting the whole point into a segment, we put part of it into one segment and part of it into another!

This is further explained on page 18 of the *Experiencing SAX: a novel symbolic representation of time series* paper. As stated in the paper, if we cannot divide the sliding window length by the number of segments, we can use a part of a point in a segment and a part of a point in another segment. We do that for points that are between two segments and not for any random points. This can be explained using an example. Imagine we have a time series such as $T = \{t_1, t_2, t_3, t_4, t_5, t_6, t_7, t_8, t_9, t_{10}\}$. For the S_1 segment, we take the values of $\{t_1, t_2, t_3\}$ and one-third of the value of t_4. For the S_2 segment, we take the values of $\{t_5, t_6\}$ and two-thirds of the values of t_4 and t_7. For the S_3 segment, we take the values of $\{t_8, t_9, t_{10}\}$ and the third of the value of t_7 that we have not used so far. This is also explained in *Figure 2.4*:

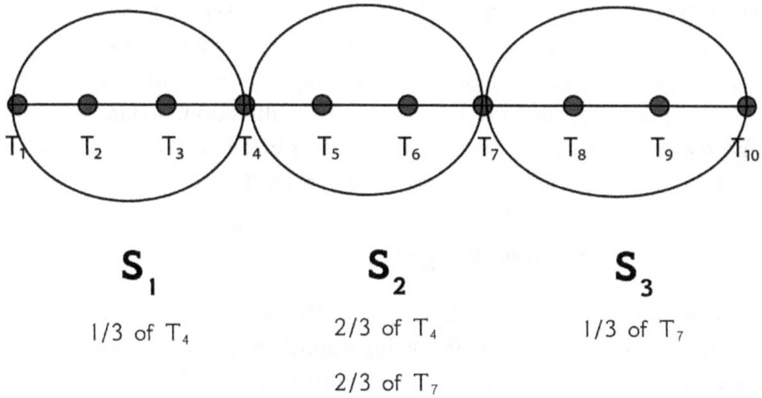

Figure 2.4 – Dividing 10 data points into 3 segments

Put simply, this is a convention decided by the creators of SAX that applies to all cases where we cannot perfectly divide the number of elements by that of the segments.

In this book, we will not deal with that case. The sliding window size, which is the length of the generated subsequences, and the number of segments are both part of a perfect division, with a remainder of 0. This simplification does not change the way SAX works, but it makes our lives a little easier.

The subject of the next subsection is how to go from higher cardinalities to lower ones without doing every computation.

Reducing the cardinality of a SAX representation

The knowledge gained from this subsection will be applicable when we discuss the iSAX index. However, as what you will learn is directly related to SAX, we have decided to discuss it here first.

Imagine that we have a SAX representation at a given cardinality and that we want to reduce the cardinality. Is that possible? Can we do this without calculating everything from scratch? The answer is simple – this can be done by ignoring trailing bits. Given a binary value of 10,100, the first trailing bit is 0, then the next trailing bit is 0, then 1, and so on. So, we start from the bits at the end, and we remove them one by one.

As most of you, including me when I first read about it, might find this unclear, let me show you some practical examples. Let us take the following two SAX representations from the *How to manually find the SAX representation of a subsequence* subsection of this chapter – [00, 11] and [010, 110, 100, 010]. To convert [00, 11] into the cardinality of 2, we must just delete the digits at the end of each SAX word. So, the new version of [00, 11] will be [0, 1]. Similarly, [010, 110, 100, 010] is going to be [01, 11, 10, 01] for the cardinality of 4 and [0, 1, 1, 0] for the cardinality of 2.

So, from a higher cardinality – a cardinality with more digits – we can go to a lower cardinality by *subtracting the appropriate number of digits from the right side* of one or more segments (the trailing bits). Can we go in the opposite direction? Not without losing accuracy, but that would still be better than nothing. However, generally, we don't go in the opposite direction. So far, we know the theory regarding the SAX representation. The section that follows briefly explains the basics of Python packages and shows the development of our own package, named sax.

Developing a Python package

In this section, we describe the process of developing a Python package that calculates the SAX representation of a subsequence. Apart from this being a good programming exercise, the package is going to be enriched in the chapters that follow when we create the iSAX index.

We will begin by explaining the basics of Python packages.

The basics of Python packages

I am not a Python expert, and the presented information is far from complete. However, it covers the required knowledge regarding Python packages.

In all but the latest Python versions, we used to need a file named __init__.py inside the directory of every Python package. Its purpose is to perform initialization actions and imports, as well as define variables. Although this is not the case with most recent Python versions, our packages will still have a __init__.py file in them. The good thing is that it is allowed to be empty if you have nothing to put into it. There is a link at the end of the chapter to the official Python documentation regarding packages, regular packages, and namespace packages, where the use of __init__.py is explained in more detail.

The next subsection discusses the details of the Python package that we are going to develop.

The SAX Python package

The code of the sax Python package is included in a directory named sax. The contents of the sax directory are presented with the help of the tree(1) command, which you might need to install on your own:

```
$ tree sax
sax
├── SAXalphabet
├── __init__.py
├── __pycache__
│   __init__.cpython-310.pyc
│   sax.cpython-310.pyc
│   tools.cpython-310.pyc
│   variables.cpython-310.pyc
├── sax.py
├── tools.py
└── variables.py

2 directories, 9 files
```

The __pycache__ directory is automatically generated by Python once you begin using the Python package and contains precompiled bytecode Python code. You can completely ignore that directory.

Let us begin by showing the contents of sax.py, which is going to be presented in multiple code chunks.

First, we have the `import` section and the implementation of the `normalize()` function, which normalizes a NumPy array:

```
import numpy as np
from scipy.stats import norm

from sax import tools

import sys
sys.path.insert(0,'..')

def normalize(x):
    eps = 1e-6
    mu = np.mean(x)
    std = np.std(x)
    if std < eps:
            return np.zeros(shape=x.shape)
    else:
            return (x-mu)/std
```

After that, we have the implementation of the `createPAA()` function, which returns the SAX representation of a time series, given the cardinality and the segments:

```
def createPAA(ts, cardinality, segments):
    SAXword = ""
    ts_norm = normalize(ts)
    segment_size = len(ts_norm) // segments
    mValue = 0
    for I in range(segments):
            ts_segment = ts_norm[segment_size * i :(i+1) * segment_
size]
            mValue = meanValue(ts_segment)
            index = getIndex(mValue, cardinality)
            SAXword += str(index) +""""
```

Python uses the double slash `//` operator to perform floor division. What the `//` operator does is divide the first number by the second number before rounding the result down to the nearest integer – this is used for the `segment_size` variable.

The rest of the code is about specifying the correct index numbers when working with the given time series (or subsequence). Hence, the `for` loop is used to process the entire time series (or subsequence) based on the segments value.

Next, we have the implementation of a function that computes the mean value of a NumPy array:

```
def meanValue(ts_segment):
    sum = 0
    for i in range(len(ts_segment)):
        sum += ts_segment[i]
    mean_value = sum / len(ts_segment)
    return mean_value
```

Finally, we have the function that returns the SAX value of a SAX word, given its mean value and its cardinality. Remember that we calculate the mean value of each SAX word separately in the `createPAA()` function:

```
def getIndex(mValue, cardinality):
    index = 0
    # With cardinality we get cardinality + 1
    bPoints = tools.breakpoints(cardinality-1)

    while mValue < float(bPoints[index]):
        if index == len(bPoints)-- 1:
                # This means that index should be advanced
                # before breaking out of the while loop
                index += 1
                break
        else:
                index += 1

    digits = tools.power_of_two(cardinality)
    # Inverse the result
    inverse_s = ""
    for i in binary_index:
        if i == '0':
                inverse_s += '1'
        else:
                inverse_s += '0'

    return inverse_s
```

The previous code computes the SAX value of a SAX word using its mean value. It iteratively visits the breakpoints, from the lowest value to the biggest, up to the point that the mean value exceeds the current breakpoint. This way, we find the index of the SAX word (mean value) in the list of breakpoints.

Now, let us discuss a tricky point, which has to do with the last statements that reverse the SAX word. This mainly has to do with whether we begin counting from the top or the bottom of the different areas that the breakpoints create. All ways are equivalent – we just decided to go that way. This is because a previous implementation of SAX used that order, and we wanted to make sure that we created the same results for testing reasons. If you want to alter that functionality, you just have to remove the last for loop.

As you saw at the beginning of this section, the sax package is composed of three Python files, not just the one that we just presented. So, we will present the remaining two files.

First, we will present the contents of variables.py:

```
# This file includes all variables for the sax package

maximumCardinality = 32

# Where to find the breakpoints file
# In this case, in the current directory
breakpointsFile =""SAXalphabe""

# Sliding window size
slidingWindowSize = 16

# Segments
segments = 0

# Breakpoints in breakpointsFile
elements ="""

# Floating point precision
precision = 5
```

You might wonder what the main reason is for having such a file. The answer is that we need to have a place to keep our global parameters and options, and having a separate file for that is a perfect solution. This will make much more sense when the code becomes longer and more complex.

Second, we present the code in tools.py:

```
import os
import numpy as np
import sys
from sax import variables

breakpointsFile = variables.breakpointsFile
maxCard = variables.maximumCardinality
```

Here, we reference two variables from the `variable.py` file, which are `variables.breakpointsFile` and `variables.maximumCardinality`:

```
def power_of_two(n):
    power = 1
    while n/2 != 1:
            # Not a power of 2
            if n % 2 == 1:
                    return -1

            n = n / 2
            power += 1

    return power
```

This is a helper function that we use when we want to make sure that a value is a power of 2:

```
def load_sax_alphabet():
    path = os.path.dirname(__file__)

    file_variable = open(path +"""" + breakpointsFile)
    variables.elements = file_variable.readlines()

def breakpoints(cardinality):
    if variables.elements =="""":
            load_sax_alphabet()

    myLine = variables.elements[cardinality-- 1].rstrip()
    elements = myLine.split'''')
    elements.reverse()
    return elements
```

The `load_sax_alphabet()` function loads the contents of the file with the definitions of breakpoints and assigns them to the `variables.elements` variable. The `breakpoints()` function returns the breakpoint values when given the cardinality.

As you can see, the code of the entire package is relatively short, which is a good thing.

In this section, we developed a Python package to compute SAX representations. In the next section, we are going to begin working with the `sax` package.

Working with the SAX package

Now that we have the SAX package at hand, it is time to use it by developing various utilities, starting with a utility that computes the SAX representations of the subsequences of a time series.

Computing the SAX representations of the subsequences of a time series

In this subsection, we will develop a utility that computes the SAX representations for all the subsequences of a time series and also presents their normalized forms. The name of the utility is `ts2PAA.py` and contains the following code:

```python
#!/usr/bin/env python3

import sys
import numpy as np
import pandas as pd

from sax import sax

def main():
    if len(sys.argv) != 5:
        print("TS1 sliding_window cardinality segments")
        sys.exit()

    file = sys.argv[1]
    sliding = int(sys.argv[2])
    cardinality = int(sys.argv[3])
    segments = int(sys.argv[4])

    if sliding % segments != 0:
        print("sliding MODULO segments != 0...")
        sys.exit()

    if sliding <= 0:
        print("Sliding value is not allowed:", sliding)
        sys.exit()

    if cardinality <= 0:
        print("Cardinality Value is not allowed:", cardinality)
        sys.exit()

    # Read Sequence as Pandas
```

```
    ts = pd.read_csv(file, names=['values'], compression='gzip')

    # Convert to NParray
    ts_numpy = ts.to_numpy()
    length = len(ts_numpy)

    PAA_representations = []
    # Split sequence into subsequences
    for i in range(length - sliding + 1):
            t1_temp = ts_numpy[i:i+sliding]
            # Generate SAX for each subsequence
            tempSAXword = sax.createPAA(t1_temp, cardinality, segments)
            SAXword = tempSAXword.split("_")[:-1]
            print(SAXword, end = ' ')
            PAA_representations.append(SAXword)

            print("[", end = ' ')
            for i in t1_temp.tolist():
                    for k in i:
                            print("%.2f" % k, end = ' ')
            print("]", end = ' ')

            print("[", end = ' ')
            for i in sax.normalize(t1_temp).tolist():
                    for k in i:
                            print("%.2f" % k, end = ' ')
            print("]")


if __name__ == '__main__':
    main()
```

The ts2PAA.py script takes a time series, breaks it into subsequences, and computes the normalized version of each subsequence using sax.normalize().

The output of ts2PAA.py is as follows (some output is omitted for brevity):

```
$ ./ts2PAA.py ts1.gz 8 4 2
['01', '10'] [ 5.22 23.44 14.14 6.75 4.31 27.94 6.61 21.73 ] [ -0.97
1.10 0.04 -0.80 -1.07 1.61 -0.81 0.90 ]
['01', '10'] [ 23.44 14.14 6.75 4.31 27.94 6.61 21.73 11.43 ] [ 1.07
-0.05 -0.94 -1.24 1.62 -0.96 0.87 -0.38 ]
['10', '01'] [ 14.14 6.75 4.31 27.94 6.61 21.73 11.43 7.15 ] [ 0.21
-0.73 -1.05 1.97 -0.75 1.18 -0.14 -0.68 ]
['01', '10'] [ 6.75 4.31 27.94 6.61 21.73 11.43 7.15 15.85 ] [ -0.76
```

```
-1.07 1.93 -0.77 1.14 -0.16 -0.70 0.40 ]
['01', '10'] [ 4.31 27.94 6.61 21.73 11.43 7.15 15.85 29.96 ] [ -1.22
1.32 -0.97 0.66 -0.45 -0.91 0.02 1.54 ]
['10', '01'] [ 27.94 6.61 21.73 11.43 7.15 15.85 29.96 6.00 ] [ 1.34
-1.02 0.65 -0.49 -0.96 0.00 1.56 -1.08 ]
. . .
```

The previous output shows the SAX representation, the original subsequence, and the normalized version of the subsequence for all the subsequences of a time series. Each subsequence is on a separate line.

> **Using Python packages**
>
> Most of the chapters that follow will need the SAX package we developed here. For reasons of simplicity, we will copy the SAX package implementation into all directories that use that package. This might not be the best practice on production systems where we want a single copy of each software or package, but it is the best practice when learning and experimenting.

So far, we have learned how to use the basic functionality of the sax package.

The next section presents a utility that counts the SAX representations of the subsequences of a time series and prints the results.

Counting the SAX representations of a time series

This section of the chapter presents a utility that counts the SAX representations of a time series. The Python data structure behind the logic of the utility is a dictionary, where the keys are the SAX representations converted into strings and the values are integers.

The code for counting.py is as follows:

```
#!/usr/bin/env python3

import sys
import pandas as pd
from sax import sax

def main():
    if len(sys.argv) != 5:
        print("TS1 sliding_window cardinality segments")
        print("Suggestion: The window be a power of 2.")
        print("The cardinality SHOULD be a power of 2.")
        sys.exit()

    file = sys.argv[1]
    sliding = int(sys.argv[2])
```

```
        cardinality = int(sys.argv[3])
        segments = int(sys.argv[4])

        if sliding % segments != 0:
                print("sliding MODULO segments != 0...")
                sys.exit()

        if sliding <= 0:
                print("Sliding value is not allowed:", sliding)
                sys.exit()

        if cardinality <= 0:
                print("Cardinality Value is not allowed:", cardinality)
                sys.exit()

        ts = pd.read_csv(file, names=['values'], compression='gzip')
        ts_numpy = ts.to_numpy()
        length = len(ts_numpy)

        KEYS = {}
        for i in range(length - sliding + 1):
                t1_temp = ts_numpy[i:i+sliding]
                # Generate SAX for each subsequence
                tempSAXword = sax.createPAA(t1_temp, cardinality, segments)
                tempSAXword = tempSAXword[:-1]

                if KEYS.get(tempSAXword) == None:
                        KEYS[tempSAXword] = 1
                else:
                        KEYS[tempSAXword] = KEYS[tempSAXword] + 1

        for k in KEYS.keys():
                print(k, ":", KEYS[k])

if __name__ == '__main__':
    main()
```

The `for` loop splits the time series into subsequences and computes the SAX representation of each subsequence using `sax.createPAA()`, before updating the relevant counter in the KEYS dictionary. The `tempSAXword = tempSAXword[:-1]` statement removes an unneeded underscore character from the SAX representation. Finally, we print the content of the KEYS dictionary.

The output of `counting.py` should be similar to the following:

```
$ ./counting.py ts1.gz 4 4 2
10_01 : 18
11_00 : 8
01_10 : 14
00_11 : 7
```

What does this output tell us?

For a time series with 50 elements (`ts1.gz`) and a sliding window size of 4, there exist 18 subsequences with the `10_01` SAX representation, 8 subsequences with the `11_00` SAX representation, 14 subsequences with the `01_10` SAX representation, and 7 subsequences with the `00_11` SAX representation. For easier comparison, and to be able to use a SAX representation as a key to a dictionary, we convert `[01 10]` into the `01_10` string, `[11 00]` into `11_00`, and so on.

> **How many subsequences does a time series have?**
>
> Keep in mind that given a time series with n elements and a sliding window size of w, the total number of subsequences is `n - w + 1`.

`counting.py` can be used for many practical tasks and will be updated in *Chapter 3*.

The next section discusses a handy Python package that can help us learn more about processing our time series from a statistical point of view.

The tsfresh Python package

This is a bonus section not directly related to the subject of the book, but it is helpful, nonetheless. It is about a handy Python package called `tsfresh`, which can give you a good overview of your time series from a statistical perspective. We are not going to present all the capabilities of `tsfresh`, just the ones that you can easily use to get information about your time series data – at this point, you might need to install `tsfresh` on your machine. Keep in mind that the `tsfresh` package has lots of package dependencies.

So, we are going to compute the following properties of a dataset – in this case, a time series:

- **Mean value**: The mean value of a dataset is the summary of all the values divided by the number of values.

- **Standard deviation**: The standard deviation of a dataset measures the amount of variation in it. There is a formula to calculate the standard deviation, but we usually compute it using a function from a Python package.

- **Skewness**: The skewness of a dataset is a measure of the asymmetry in it. The value of skewness can be positive, negative, zero, or undefined.

- **Kurtosis**: The kurtosis of a dataset is a measure of the tailedness of a dataset. In more mathematical terms, kurtosis measures the heaviness of the tail of a distribution compared to a normal distribution.

All these quantities will make much more sense once you plot your data, which is left as an exercise for you; otherwise, they will be just numbers. So, now that we know some basic statistic terms, let us present a Python script that calculates all these quantities for a time series.

The Python code for `using_tsfresh.py` is as follows:

```python
#!/usr/bin/env python3

import sys
import pandas as pd
import tsfresh

def main():
    if len(sys.argv) != 2:
        print("TS")
        sys.exit()

    TS1 = sys.argv[1]
    ts1Temp = pd.read_csv(TS1, compression='gzip')
    ta = ts1Temp.to_numpy()
    ta = ta.reshape(len(ta))

    # Mean value
    meanValue = tsfresh.feature_extraction.feature_calculators.mean(ta)
    print("Mean value:\t\t", meanValue)

    # Standard deviation
    stdDev = tsfresh.feature_extraction.feature_calculators.standard_deviation(ta)
    print("Standard deviation:\t", stdDev)

    # Skewness
    skewness = tsfresh.feature_extraction.feature_calculators.skewness(ta)
    print("Skewness:\t\t", skewness)

    # Kurtosis
```

```
        kurtosis = tsfresh.feature_extraction.feature_calculators.
    kurtosis(ta)
        print("Kurtosis:\t\t", kurtosis)

    if __name__ == '__main__':
        main()
```

The output of using_tsfresh.py when processing ts1.gz should look similar to the following:

```
$ ./using_tsfresh.py ts1.gz
Mean value:   15.706410001204729
Standard deviation:   8.325017802111901
Skewness:        0.008971113265160474
Kurtosis:       -1.2750042973761417
```

The tsfresh package can do many more things; we have just presented the tip of the iceberg of the capabilities of tsfresh.

The next section is about creating a histogram of a time series.

Creating a histogram of a time series

This is another bonus section, where we will illustrate how to create a histogram of a time series to get a better overview of its values.

A **histogram**, which looks a lot like a bar chart, defines buckets (bins) and counts the number of values that fall into each bin. Strictly speaking, a histogram allows you to understand your data by creating a plot of the **distribution of values**. You can see the maximum and the minimum values, as well as find out data patterns, just by looking at a histogram.

The Python code for histogram.py is as follows:

```
#!/usr/bin/env python3

import sys
import pandas as pd
import matplotlib.pyplot as plt
import numpy as np
import math
import os

if len(sys.argv) != 2:
    print("TS1")
    sys.exit()
```

```
TS1 = sys.argv[1]

ts1Temp = pd.read_csv(TS1, compression='gzip')
ta = ts1Temp.to_numpy()
ta = ta.reshape(len(ta))

min = np.min(ta)
max = np.max(ta)

plt.style.use('Solarize_Light2')
bins = np.linspace(min, max, 2 * abs(math.floor(max) + 1))

plt.hist([ta], bins, label=[os.path.basename(TS1)])
plt.legend(loc='upper right')

plt.show()
```

The third argument of the np.linespace() function helps us define the number of bins the histogram has. The first parameter is the minimum value, and the second parameter is the maximum value of the presented samples. This script does not save its output in a file but, instead, opens a window on your GUI to display the output. The plt.hist() function creates the histogram, whereas the plt.legend() function puts the legend in the output.

A sample output of histogram.py can be seen in *Figure 2.5*:

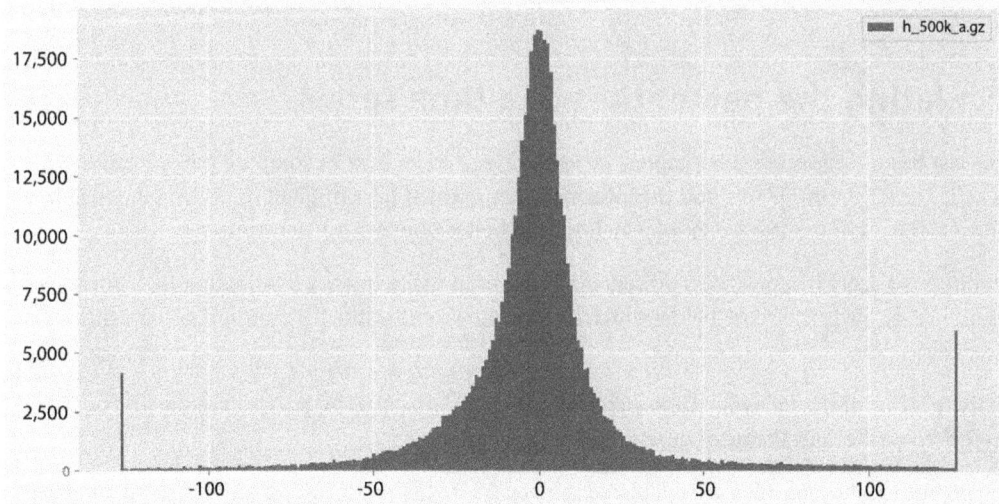

Figure 2.5 – A sample histogram

A different sample output from `histogram.py` can be seen in *Figure 2.6*:

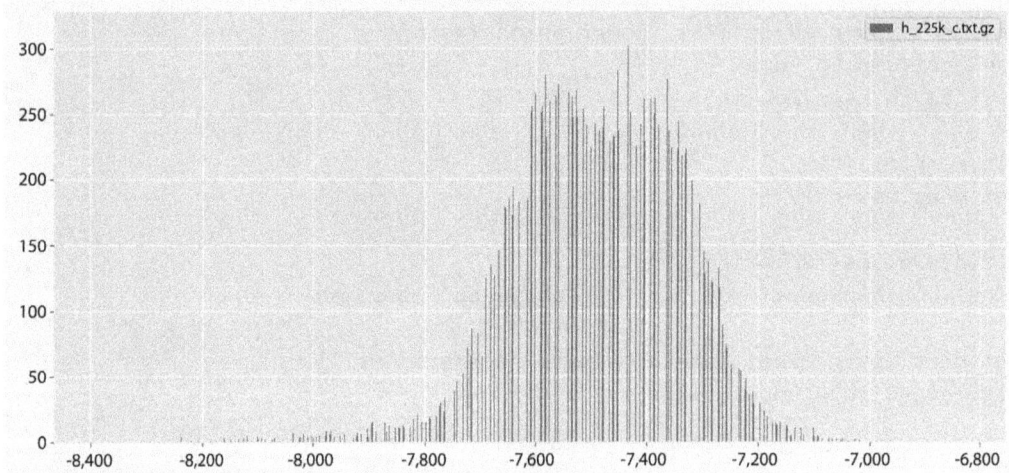

Figure 2.6 – A sample histogram

So, what is the difference between the histograms in *Figure 2.5* and *Figure 2.6*? There exist many differences, including the fact that the histogram in *Figure 2.5* does not have empty bins and it contains both negative and positive values. On the other hand, the histogram in *Figure 2.6* contains negative values only that are far away from 0.

Now that we know about histograms, let us learn about another interesting statistical quantity – *percentiles*.

Calculating the percentiles of a time series

In this last bonus section of this chapter, we are going to learn how to compute the percentiles of a time series or a list (and if you find the information presented here difficult to understand, feel free to skip it). The main usage of such information is to better understand your time series data.

A *percentile* is a score where a given percentage of scores in the frequency distribution falls. Therefore, the 20th percentile is the score below which 20% of the scores of the distribution of the values of a dataset falls.

A **quartile** is one of the following three percentiles – 25%, 50%, or 75%. So, we have the first quartile, the second quartile, and the third quartile, respectively.

Both percentiles and quartiles are calculated in datasets sorted in ascending order. Even if you have not sorted that dataset, the relevant NumPy function, which is called `quantile()`, does that behind the scenes.

The Python code of `percentiles.py` is as follows:

```python
#!/usr/bin/env python3

import sys
import pandas as pd
import numpy as np

def main():
    if len(sys.argv) != 2:
        print("TS")
        sys.exit()

    F = sys.argv[1]
    ts = pd.read_csv(F, compression='gzip')
    ta = ts.to_numpy()
    ta = ta.reshape(len(ta))

    per01 = round(np.quantile(ta, .01), 5)
    per25 = round(np.quantile(ta, .25), 5)
    per75 = round(np.quantile(ta, .75), 5)

    print("Percentile 1%:", per01, "Percentile 25%:", per25,
"Percentile 75%:", per75)

if __name__ == '__main__':
    main()
```

All the work is done by the `quantile()` function of the NumPy package. Among other things, `quantile()` appropriately arranges its elements before performing any calculations. We do not know what happens internally, but most likely, `quantile()` sorts its input in ascending order.

The first parameter of `quantile()` is the NumPy array, and its second parameter is the percentage (percentile) that interests us. A 25% percentage is equal to the first quantile, a 50% percentage is equal to the second quantile, and a 75% percentage is equal to the third quantile. A 1% percentage is equal to the 1% percentile, and so on.

The output of `percentiles.py` is as follows:

```
$ ./percentiles.py ts1.gz
Percentile 1%: 1.57925 Percentile 25%: 7.15484 Percentile 75%: 23.2298
```

Summary

This chapter included the theory behind and practical implementation of SAX and an understanding of a time series from a statistical viewpoint. As the iSAX index construction is based on the SAX representation, we cannot construct an iSAX index without computing SAX representations.

Before you begin reading *Chapter 3*, please make sure that you know how to calculate the SAX representation of a time series or a subsequence, given the sliding window size, the number of segments, and the cardinality.

The next chapter contains the theory related to the iSAX index, shows you how to manually construct an iSAX index (which you will find very entertaining), and includes the development of some handy utilities.

Useful links

- About Python packages: https://docs.python.org/3/reference/import.html
- The tsfresh package: https://pypi.org/project/tsfresh/
- The documentation for the tsfresh package can be found at https://tsfresh.readthedocs.io/en/latest/
- The scipy package: https://pypi.org/project/scipy/ and https://scipy.org/
- Normalization: https://en.wikipedia.org/wiki/Normalization_(statistics)
- Histogram: https://en.wikipedia.org/wiki/Histogram
- Percentile: https://en.wikipedia.org/wiki/Percentile
- Normal distribution: https://en.wikipedia.org/wiki/Normal_distribution

Exercises

Try to solve the following exercises in Python:

- Divide *by hand* the y axis for the $16 = 2^4$ cardinality. Did you divide it into 16 areas or 17 areas? How many breakpoints did you use?
- Divide *by hand* the y axis for the $64 = 2^6$ cardinality. Did you divide it into 64 areas?
- Use the cardinality.py utility to plot the breakpoints of the $16 = 2^4$ cardinality.
- Use the cardinality.py utility to plot the breakpoints of the $128 = 2^7$ cardinality.
- Find the SAX representation of the {0, 2, -1, 2, 3, 4, -2, 4} subsequence using 4 segments and a cardinality of 4 (2^2). Do not forget to normalize it first.

- Find the SAX representation of the $\{0, 2, -1, 2, 3, 4, -2, 4\}$ subsequence using 2 segments and a cardinality of 2 (2^1). Do not forget to normalize it first.

- Find the SAX representation of the $\{0, 2, -1, 2, 3, 1, -2, -4\}$ subsequence using 4 segments and a cardinality of 2.

- Given the $\{0, -1, 1.5, -1.5, 0, 1, 0\}$ time series and a sliding window size of 4, find the SAX representation of all its subsequences using 2 segments and a cardinality of 2.

- Create a synthetic time series and process it using using_tsfresh.py.

- Create a synthetic time series with 1,000 elements and process it using histogram.py.

- Create a synthetic time series with 5,000 elements and process it using histogram.py.

- Create a synthetic time series with 10,000 elements and process it using counting.py.

- Create a synthetic time series with 100 elements and process it using percentiles.py.

- Create a synthetic dataset with 100 elements and examine it using counting.py.

- Modify histogram.py to save its graphical output in a PNG file.

- Plot a time series using histogram.py and then process it using using_tsfresh.py

3
iSAX – The Required Theory

Now that we know all about SAX, including normalization and computing the SAX representation of a subsequence, it is time to learn the theory behind the iSAX index, which, at the time of writing, is considered one of the best time-series indexes. Improved versions of iSAX that make iSAX faster and more compact exist, but the core ideas remain the same.

As you might have guessed from its name, iSAX depends on SAX in some way. Put simply, **the keys to every iSAX index are SAX representations**. Therefore, searching in an iSAX index depends on SAX representations.

At this point, I believe it would be good to provide more information about iSAX to help you while reading this chapter. An **iSAX index** is a tree-like structure where the root, and only the root, can have multiple children, and all the children of the root are binary trees underneath. Additionally, to create an iSAX index, we need a time series and a threshold value, which is the maximum number of subsequences that a leaf node (a terminal node in iSAX terminology) can store, as well as a segment value and a cardinality value. The last two parameters are related to the SAX representation. All of this is explained in more detail in this chapter, but it's good to have the big picture in mind as early on as possible.

Additionally, in this chapter, we are going to manually construct a small iSAX index step by step to better understand the process, using lots of visualizations. Please make sure that you carry out this process on your own after you finish reading this chapter.

In this chapter, we are going to cover the following main topics:

- Background information
- Understanding how iSAX works
- How iSAX is constructed
- Manually constructing an iSAX index
- Updating the counting.py utility

Technical requirements

The GitHub repository for this book can be found at `https://github.com/PacktPublishing/Time-Series-Indexing`. The code for each chapter is in its own directory. Therefore, the code for *Chapter 3* can be found in the `ch03` folder.

Background information

In this first section, we are going to learn about the basic definitions and concepts related to iSAX. But first, we are going to mention the research paper that describes the operation of iSAX. iSAX and its operation are described in *iSAX: disk-aware mining and indexing of massive time series datasets*, which was written by Jin Shieh and Eamonn Keogh. You do not have to read this paper from start to finish but, as we mentioned for the SAX research paper, it would benefit you to read its abstract and introduction section.

Additionally, there have been various improvements to iSAX, mainly to make it faster, which are presented in the following research papers:

- *iSAX 2.0: Indexing and Mining One Billion Time Series*, written by Alessandro Camerra, Themis Palpanas, Jin Shieh, and Eamonn Keogh

- *Beyond one billion time series: Indexing and mining very large time series collections with iSAX2+*, written by Alessandro Camerra, Jin Shieh, Themis Palpanas, Thanawin Rakthanmanon, and Eamonn Keogh

- *DPiSAX: Massively Distributed Partitioned iSAX*, written by Djamel-Edine Yagoubi, Reza Akbarinia, Florent Masseglia, and Themis Palpanas

- *Evolution of a Data Series Index: The iSAX Family of Data Series Indexes: iSAX, iSAX2.0, iSAX2+, ADS, ADS+, ADS-Full, ParIS, ParIS+, MESSI, DPiSAX, ULISSE, Coconut-Trie/Tree, Coconut-LSM*, written by Themis Palpanas

We are not going to deal with the aforementioned research papers in this book as we are working with the initial version of the iSAX index. You do not have to read all these papers for the purposes of this book, but it would be a great exercise to have a look through them when you feel comfortable with iSAX.

As iSAX is a tree, the next subsection presents basic information about trees and binary trees.

Trees and binary trees

Let us begin by explaining what a directed graph is. A **directed graph** is a graph where the edges have a direction associated with them. A **directed acyclic graph** (**DAG**) is a directed graph with no cycles in it.

In computer science, a **tree** is a DAG data structure that satisfies the following three principles:

- It has a root node that is the entry point to the tree

- Every vertex, except the root, has one and only one entry point

- There is a path that connects the root to each vertex

As already stated, the **root of a tree** is the first node of the tree. Each node can be connected to one or more nodes depending on the tree type. If each node leads to one and only one node, then the tree becomes a linked list! A leaf node is a node without any children. Leaves are also called external nodes, whereas a node with at least one child is called an internal node.

A **binary tree** is a tree where underneath each node, there are at most two more nodes. *At most* means that it can be connected to one, two, or no other nodes. The **depth**, which is also called the **height**, of a tree is defined as the longest path from the root node to a leaf, whereas the depth of a node is the number of edges from the node to the root node of the tree.

Keep in mind that if you create two binary trees using the same set of elements added in a different order, you are going to get two completely different trees. The simplest way to do that is by starting from a different root node. So, we do not know in advance the final shape of a tree.

A tree has internal nodes and leaf nodes. In iSAX terminology, these are called **inner nodes** and **terminal nodes**, respectively. *Inner nodes* have a parent and at least one child and *terminal nodes* have a parent node but no children of their own.

A binary tree is considered a **balanced tree** when the difference between the longest length from the root node to a leaf node and the shortest such length is 0 or 1. An **unbalanced tree** is a tree that is not balanced. Although the iSAX index can be characterized by its length, there is no point in finding out whether an iSAX index is balanced or not because this is not how indexes work and this mainly has to do with the data. However, having balanced trees is a good thing.

Therefore, although iSAX indexes are trees, we cannot apply all the binary tree rules to an iSAX index. However, it is good to know all these details to have a better understanding of why a particular iSAX index might be faster or slower than others.

Balanced binary trees

If a binary tree is balanced, its search, insert, and delete operations take about `log (n)` steps, where n is the total number of elements in the tree. Additionally, the height of a balanced binary tree is approximately $log_2 n$, which means that a balanced tree with 10,000 elements has a height of 14, which is remarkably small. The height of a balanced tree with 100,000 elements will be 17 and the height of a balanced tree with 1,000,000 elements will be 20! In other words, putting a significantly large number of elements into a balanced binary tree does not change the search speed of the tree extremely. Put differently, you can reach any node in a balanced tree with 1,000,000 nodes in at most 20 steps!

Figure 3.1 shows a small iSAX index, which allows us to learn more about iSAX indexes without getting lost in the details. We are going to learn more about the details of an iSAX index starting in the next section.

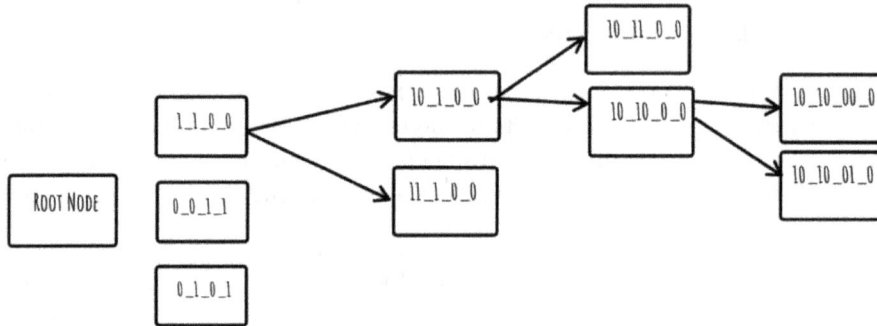

Figure 3.1 – A small iSAX index

So, what do we have here and what does it tell us?

As each node of the iSAX index has four SAX words, we know that the segment value is 4. However, we cannot be sure about the cardinality as cardinalities begin from the smallest value and are increased when needed, that is, when there is a split. However, even on splits, only a single SAX word is promoted to a higher cardinality value. The root of this particular iSAX index has three children – we assume that the remaining children of the root node are empty. The terminal nodes are [0 0 1 1], [0 1 0 1], [11 1 0 0], [10 11 0 0], [10 10 00 0], and [10 10 01 0], and the inner nodes are [1 1 0 0], [10 1 0 0], and [10 10 0 0]. We cannot make any assumptions about the threshold value.

Chapter 6 is going to present techniques for visualizing iSAX indexes, which can be very handy when we are working with big iSAX indexes and want to have an overview of the indexes.

The next section will provide more details about iSAX.

Understanding how iSAX works

In this section, we will discuss the way iSAX works, which includes the construction phase as well as its usage and parameters. Apart from the required theory, we will also present a handy command-line utility that helps you understand how many subsequences an iSAX index can have, given its parameters.

> **How iSAX and SAX are related**
>
> The way iSAX and the SAX representation are related is simple. The keys on all the nodes of an iSAX index, apart from the root node, which has no key, are all SAX representations. So, a big part of building and searching an iSAX index is based on SAX.

We do not delete or update elements in an iSAX index not because it is not possible but because this is not how an iSAX index works.

It is now time to discuss the parameters of an iSAX index because the construction of an iSAX depends on them.

The cardinality parameter

The **cardinality** parameter is the same as in SAX, but it has some small differences for the sake of efficiency. The first difference is that iSAX nodes can have multiple cardinalities. So, we might have a node with the [1 000 0 11] SAX representation and another one with the [10001 00 0 1] SAX representation. The second difference is that we begin with a cardinality value of 2. This means that all the children of the root node have a cardinality value of 2 in all their SAX words.

The segments parameter

The **segments** parameter of an iSAX index works in completely the same way as in every SAX representation and defines *the number of SAX words* in each SAX representation.

The threshold parameter

The **threshold** parameter is new and defines *the maximum number of subsequences* a terminal node can store. We cannot exceed that value. The threshold parameter defines when the split of a node is going to take place.

The next subsection presents a command line utility that computes the mean values of a normalized subsequence based on the number of segments because the mean values define the SAX words.

Computing the normalized mean values

This subsection presents a utility that outputs the mean values of all normalized subsequences of a time series based on the segment value. It would make no sense to count the mean values of all normalized subsequences without splitting them based on the segment value because after normalization, the mean value of a subsequence is very close to 0, which is one of the points of normalization.

We will use the meanValues.py script in this example. Its code is as follows:

```
#!/usr/bin/env python3

# Date: Monday 19 December 2022
#
```

```
# This utility outputs the mean values of all
# (NORMALIZED) subsequences of a time series

import sys
import numpy as np
import pandas as pd

sys.path.insert(0,'..')

def normalize(x):
    eps = 1e-6
    mu = np.mean(x)
    std = np.std(x)
    if std < eps:
        return np.zeros(shape=x.shape)
    else:
        return (x-mu)/std

def main():
    if len(sys.argv) != 4:
        print("Usage: TS1 sliding_window segments")
        sys.exit()

    file = sys.argv[1]
    # We prefer values which are powers of 2
    sliding = int(sys.argv[2])
    segments = int(sys.argv[3])

    if sliding <= 0:
        print("Sliding value is not allowed:", sliding)
        sys.exit()

    ts = pd.read_csv(file, names=['values'],
        compression='gzip')
    ts_numpy = ts.to_numpy()
    length = len(ts_numpy)

    splits = sliding // segments
    # Split time series into subsequences
    for i in range(length - sliding + 1):
```

```
t1_temp = ts_numpy[i:i+sliding]
normalized = normalize(t1_temp)

for s in range(segments):
    temp = normalized[splits*s:splits*(s+1)]
    mValue = np.mean(temp)
    print(round(mValue,5))
```

Based on the number of segments, the utility splits each subsequence into parts and calculates the mean value of each part. The last part of the code is as follows:

```
if __name__ == '__main__':
    main()
```

The output of meanValues.py is the following (a big part of the output is omitted for brevity):

```
$ ./meanValues.py ts1.gz 16 4
-0.33294
-0.00404
0.10926
0.22772
-0.51625
0.05592
-0.20672
. . .
```

You can save the output and process it with a utility such as histogram.py, which was presented in the previous chapter.

Figure 3.2 shows the output of histogram.py from two different time series with about 500,000 elements each when processed with meanValues.py using a sliding window size of 1024 and 4 segments. Four segments means that for each subsequence, we have to compute four mean values because we split each subsequence into four parts. Therefore, a 500,000-element time series produces 2,000,000 mean values that go into the histogram.

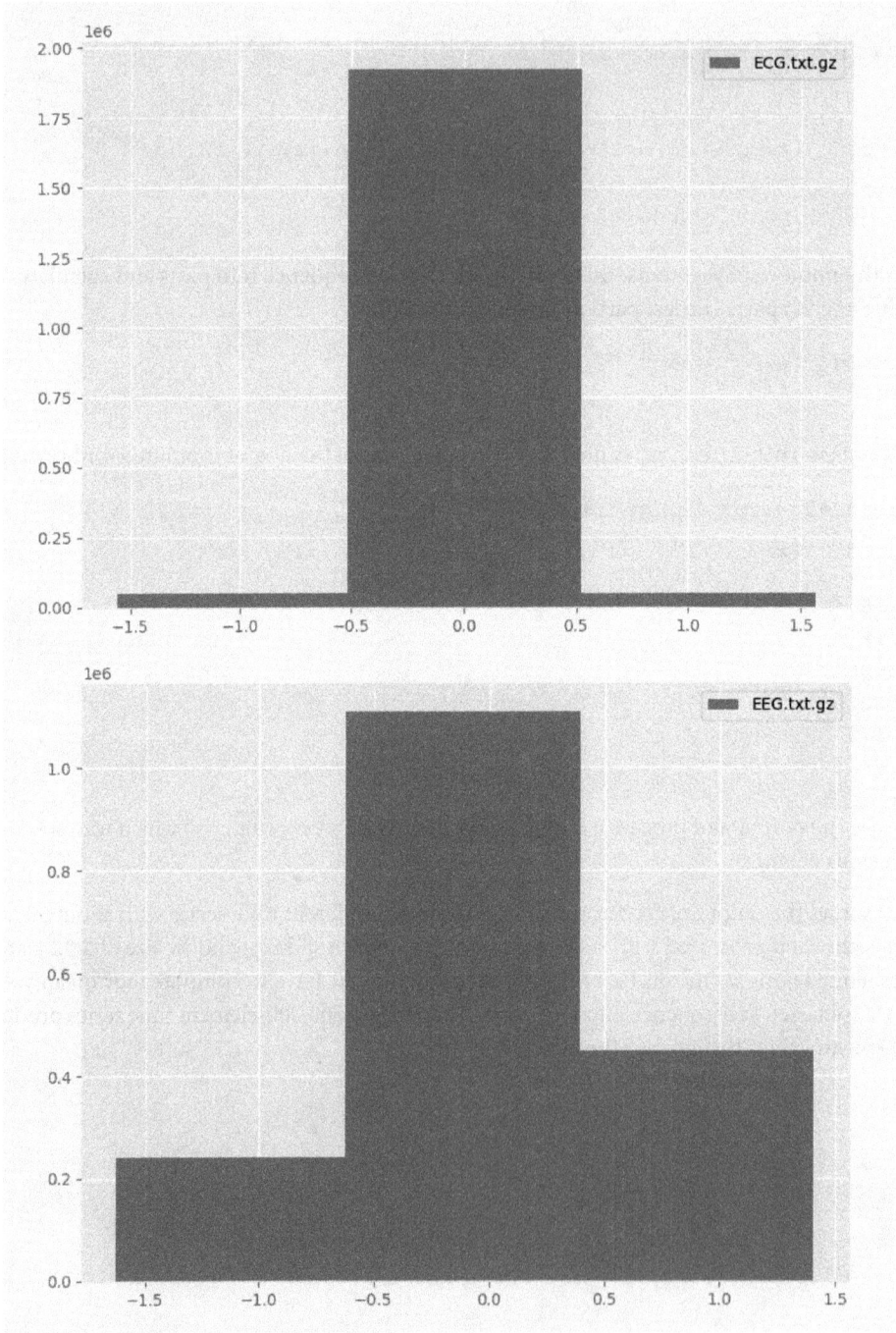

Figure 3.2 – The output of histogram.py from two time series

What does *Figure 3.2* tell us? The upper plot is of a time series named ECG and the lower histogram is of a time series named EEG. The vast majority (around 2 million values) of the mean values of the ECG time series, *and therefore the cardinality values*, falls into the -0.5 to 0.5 value range. This means that, based on the breakpoint values, many SAX words are going to be very similar to each other and therefore are going to fall into a small amount of iSAX branches, making *the iSAX index of the ECG dataset really unbalanced*. On the other hand, for the EEG dataset, the mean values are better distributed in the histogram, which means that the SAX words they are going to produce are going to be better distributed and the generated iSAX index is going to be more balanced.

If you want to find out more about the distribution of values in a histogram, you can increase the number of bins and get a more detailed output.

The next subsection presents a command line utility that computes the maximum number of subsequences an iSAX index with the given parameters can store. Keep in mind that this is an **ideal case** where the iSAX index is completely balanced and all terminal nodes have the maximum number of subsequences, which is not usually the case.

How big can an iSAX index get?

In this subsection, we are going to calculate the maximum number of subsequences an iSAX index can store in the ideal condition. As you already know, the root of an iSAX index can have more than two children, whereas all other inner nodes have one or two children, which can either be other inner nodes or terminal nodes. An inner node can have any combination of inner nodes and terminal nodes as children.

Now, let us do an exercise together: imagine that we want to find out the maximum number of nodes an iSAX with a cardinality of 4 and 2 segments can have. As we are interested in nodes, the threshold parameter does not play a key role in our discussion, so it is ignored for now. We are going to come back to the threshold parameter at the end of our discussion.

First, we are going to calculate the number of children of the root node. This only depends on the number of segments because all the children of the root have a cardinality of 2. Therefore, if we have 2 segments, the maximum number of children of the root node is 4, which is 2^2. These have the following SAX representations: [0, 0], [0, 1], [1, 0], and [1, 1]. If you are familiar with the binary system, these are all values that can be represented with 2 bits.

If we have 3 segments, the maximum number of children of the root node is 8, which is 2^3. These have the following SAX representations: [0, 0, 0], [0, 0, 1], [0, 1, 0], [0, 1, 1], [1, 0, 0], [1, 0, 1], [1, 1, 0], and [1, 1, 1]. If you are familiar with the binary system, these are all values that can be represented with 3 bits. For each of these children of the root, each SAX word can be as big as the cardinality value.

Now, back to our original problem, which is computing the maximum number of nodes an iSAX with a cardinality of 4 and 2 segments can have. As proved, when there are 2 segments, the number of children of the root node is 4. Each segment can have as many values as the cardinality, which in this case is 4. So, with 2 segments, we can have 4 times 4 possible combinations, which is 16. This is the maximum value of terminal nodes that this iSAX index can store. So, *when we have a cardinality of 4 and 2 segments, the maximum number of terminal nodes is 16.*

Now that we know the total number of terminal nodes, it is time to stop ignoring the threshold parameter and see what kind of information it can provide us. The threshold parameter can help us compute the maximum number of subsequences this iSAX index can hold in an **ideal situation**. The reason we are talking about an ideal situation is that most of the time, iSAX indexes are not balanced and there is no way of making sure that an iSAX index is going to be balanced. This is because this depends on the subsequences and the sliding window size.

Therefore, *when we have 64 terminal nodes and a threshold value of 100, the maximum number of subsequences that can be stored in an iSAX index in an ideal case is 6,400.*

With all that information in mind, we are going to develop a small Python script to do the calculations for us. The code for `maximumISAX.py` is as follows:

```
#!/usr/bin/env python3

import sys

def main():
    if len(sys.argv) != 4:
        print("cardinality segments threshold")
        print("Suggestion: The window be a power of 2.")
        print("The cardinality SHOULD be a power of 2.")
        sys.exit()

    cardinality = int(sys.argv[1])
    segments = int(sys.argv[2])
    threshold = int(sys.argv[3])

    terminalNodes = pow(cardinality, segments)
    print("Nodes:", terminalNodes)

    subsequences = terminalNodes * threshold
    print("Maximum number of subsequences:", subsequences)

if __name__ == '__main__':
    main()
```

The output of `maximumISAX.py` is the following:

```
$ ./maximumISAX.py 4 4 100
Nodes: 256
Maximum number of subsequences: 25600
```

So, the previous output tells us that an iSAX index with 4 segments, a cardinality value of 4, and a threshold value of 100 can hold up to 25600 subsequences.

If we run maximumISAX.py with the same number of segments and a different cardinality, we are going to get the following output:

```
$ ./maximumISAX.py 16 4 100
Nodes: 65536
Maximum number of subsequences: 6553600
```

So, the previous output tells us that an iSAX index with 4 segments, a cardinality value of 16, and a threshold value of 100 can hold up to 6553600 subsequences despite the fact that the root node still has 16 children.

The next subsection will discuss what happens when there is no space to add a given subsequence to an iSAX index.

What happens when there is no space left for adding more subsequences to an iSAX index?

An iSAX index can *overflow*. In practice, this means that an iSAX index might not have enough space to add more subsequences. This happens when one or more branches of an iSAX index have used the full cardinality for all segments and the threshold value has been reached in these terminal nodes. *Figure 3.3* visualizes this:

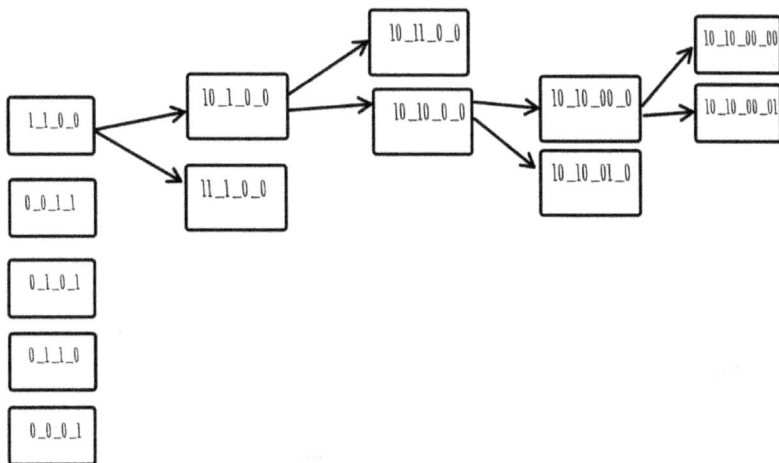

Figure 3.3 – iSAX overflow

Suppose that the nodes with SAX representations of [10 10 00 00] and [10 10 00 01] are full and are using their full cardinality. This means that we cannot promote the nodes with the [10 10 00 00] and [10 10 00 01] SAX representations anymore unless we increase one or more relevant parameters of the iSAX index. So, if we have a subsequence with a SAX representation of either [10 10 00 00] or [10 10 00 01], we do not know what to do with it, hence the overflow of the iSAX index.

In this section, we learned basic information about iSAX. The next section discusses the iSAX construction in more detail.

How iSAX is constructed

This section is going to describe the way an iSAX index is constructed. All the presented information is based on the research paper that describes iSAX. The logical steps are as follows:

- We begin with a node that is the root of the iSAX index. The root contains no actual data (subsequences) but it contains pointers to all nodes with the specified segments value and a cardinality value of 2, which is a single bit that can only have two values, 0 and 1.
- After that, we construct the children of the root node, which at this initial point are all terminal nodes without any subsequences in them.
- We now begin adding subsequences to the children of the root based on their SAX representation.
- When the threshold value of a terminal node is reached, we perform the splits, based on the specified **promotion strategy**, and we distribute the subsequences to the two newly created terminal nodes.
- The process goes on until all subsequences have been inserted into the iSAX index or there is no place left to add a subsequence.

The next section will explain how iSAX is searched in order to get a better idea of how the construction process is related to the search operation.

How iSAX is searched

The search process of an iSAX index can be described as follows:

- First, we have a subsequence that we want to look for in an iSAX index.
- Then, we must compute the SAX representation of that subsequence, using the same parameters as the iSAX index.
- After that, we use the SAX representation of the subsequence to find out the child of the root that it belongs to.

- Then, we keep searching that binary tree until we find a node with the same SAX representation. Keep in mind that we might need to reduce some of the SAX words of the subsequence based on the iSAX nodes. If such a node does not exist, then the iSAX index does not contain that subsequence.

- Last, we examine the subsequences of that node – provided that such a terminal node exists – and look for that specific subsequence.

Generally speaking, the process is the same as if we added that subsequence to the iSAX index up to the point where we found the node that it belongs to. After that, we search the subsequences of that node.

The next subsection will discuss the promotion strategy, which has to do with which SAX word is going to get promoted when a split needs to happen.

Promotion strategy

The term **promotion strategy** refers to the selection of the segment that is going to be updated to a higher cardinality when a split is about to happen. Splits happen when we are about to add a new subsequence to a terminal node that already holds a number of subsequences that is equal to the threshold value. *Only that terminal node is going to be split.*

> **The threshold value**
> The only parameter that is responsible for the splitting of a terminal node is the threshold parameter. If the threshold parameter were limitless, then all iSAX indexes would have terminal nodes with a cardinality value of 2, which are the root children.

There are two promotion strategies when splitting a node:

- **Round Robin**: The Round Robin strategy is well known in the Computer Science field. Its logic is simple: every time we need to promote, we promote the SAX word that is on the right to the SAX word that was previously promoted. When we reach the rightmost SAX word, we continue the process from the first SAX word. So, imagine we have the [0 1 0 0] SAX representation. For the first promotion that needs to take place, we promote the first SAX word and get [00 1 0 0] and [01 1 0 0]. In the next promotion, we are going to promote the second SAX word, even if this promotion takes place on a different SAX representation. So, if we need to promote [1 1 0 1], then this is going to give us [1 10 0 1] and [1 11 0 1].

- **From Left to Right**: In this strategy, we *always promote the leftmost SAX word*, if it has not reached its maximum cardinality. So, promoting [00 1 0 0] is going to give us [000 1 0 0] and [001 1 0 0], given that we have a cardinality value of at least 8. When the leftmost SAX word has reached its full cardinality, we continue with the next one to its right.

There is no right or wrong promotion strategy – use what you like best.

The next subsection presents an essential operation of iSAX, which is node splitting.

Splitting nodes

This subsection discusses the node-splitting process in more detail. Splitting happens in one case only: when a terminal node is about to have more subsequences than the ones permitted by the threshold value. In that case, **that terminal node becomes an inner node**. Then, two terminal nodes are created based on the promotion strategy, which are the children of the newly created inner node. Imagine having the iSAX index presented in *Figure 3.4*, which currently stores 10 subsequences: S_0 to S_9.

We now want to add the S_{10} subsequence to that index, but [10 10 00 0] is full:

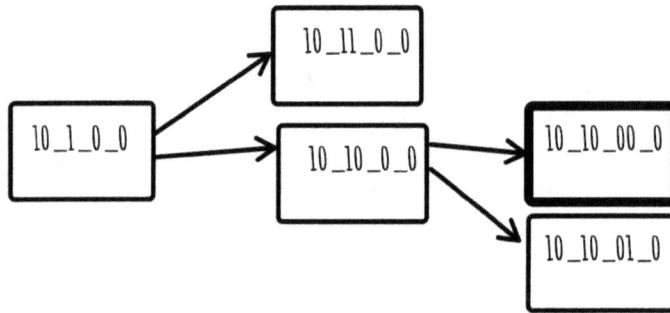

Figure 3.4 – Before node splitting

So, in order to add a subsequence with a SAX representation of [10 10 00 01], we need to split the [10 10 00 0] terminal node, which now becomes an inner node with the same SAX representation, and **create two terminal nodes** as its children. Using the *Round Robin promotion strategy*, the two new terminal nodes will have the [10 10 00 00] and [10 10 00 01] SAX representations, respectively. So, in that case, we promoted the last SAX word.

After that, we iterate all the subsequences that were previously stored in [10 10 00 0] in order to put them in either [10 10 00 00] or [10 10 00 01] based on their promoted SAX representations. We are not going to go into more detail about how the subsequences are distributed after the splitting because this is explained in detail in the *Manually constructing an iSAX index* section that follows.

> **What if splitting does not solve the issue?**
>
> In the previous example, S_{10} is going to go to the [10 10 00 01] terminal node, *provided that there is enough space* there. In the rare case where the new subsequences as well as all previously stored ones go to the same terminal node, the splitting process keeps going until the situation is resolved on its own or there is an iSAX overflow.

Figure 3.5 shows the new version of the iSAX index, after the node splitting:

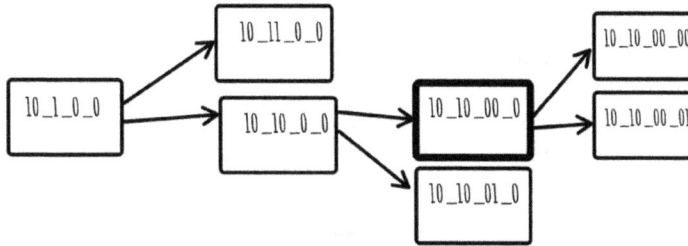

Figure 3.5 – After node splitting

In this section, we learned about the stages of the iSAX construction process. The following section will show the manual construction of an iSAX index using a real example and diagrams.

Manually constructing an iSAX index

In this section, we are going to manually create a small iSAX index. For a better understanding of the process, we are going to present all the steps and describe all the required computations.

If you recall from earlier on in this chapter, the steps for creating an iSAX index can be described as follows:

1. Separate a time series into subsequences based on the given sliding window size.

2. For each subsequence, calculate its SAX representation, based on the given parameters.

3. Begin inserting the subsequences of the time series into the iSAX index. In the beginning, all iSAX nodes are terminal nodes, apart from the root.

4. Once a terminal node is full – the threshold value has been reached – split that node by increasing the cardinality of one of its segments and create two new terminal nodes.

5. The original terminal node becomes an inner node, which is now the father of the new newly created terminal nodes.

6. Split the subsequences of the original terminal node into the two new terminal nodes based on their SAX representation.

Where does a subsequence go?

It is important to realize that a subsequence with the [00 10 10 11] SAX representation *must go* under the [0 1 1 1] child of the root. After that, the exact place (terminal node) depends on the promotion that is going to take place in the process, which can be [00 10 10 1], [00 1 1 11], [0 1 10 11], and so on. However, if we use its full SAX representation, it can only go in the [00 10 10 11] terminal node.

We are going to use a small time series for the iSAX construction process to not take too long to finish. However, the principles remain the same. We are also going to need the help of the ts2PAA. py utility from the previous chapter to get the SAX representation of each subsequence – we do not want to calculate everything manually.

As we are dealing with a small time series, we are going to use a segment value of 2, a cardinality value of 8, which means that we are going to be using 3 bits for the cardinality, and a threshold value of 15. The sliding window size is 8.

So, the output of ts2PAA.py when processing ts1.gz is as follows (we are ignoring the normalized part of the output at this point, which is omitted for brevity):

```
$ ../ch02/ts2PAA.py ts1.gz 8 8 2
[011, 100]  [5.22 23.44 14.14 6.75 4.31 27.94 6.61 21.73]
[011, 100]  [23.44 14.14 6.75 4.31 27.94 6.61 21.73 11.43]
[100, 011]  [14.14 6.75 4.31 27.94 6.61 21.73 11.43 7.15]
[011, 100]  [6.75 4.31 27.94 6.61 21.73 11.43 7.15 15.85]
[011, 100]  [4.31 27.94 6.61 21.73 11.43 7.15 15.85 29.96]
[100, 011]  [27.94 6.61 21.73 11.43 7.15 15.85 29.96 6.00]
[010, 101]  [6.61 21.73 11.43 7.15 15.85 29.96 6.00 20.74]
[011, 100]  [21.73 11.43 7.15 15.85 29.96 6.00 20.74 18.39]
[011, 100]  [11.43 7.15 15.85 29.96 6.00 20.74 18.39 23.23]
[010, 101]  [7.15 15.85 29.96 6.00 20.74 18.39 23.23 25.71]
[011, 100]  [15.85 29.96 6.00 20.74 18.39 23.23 25.71 18.74]
[011, 100]  [29.96 6.00 20.74 18.39 23.23 25.71 18.74 15.09]
[100, 011]  [6.00 20.74 18.39 23.23 25.71 18.74 15.09 1.22]
[101, 010]  [20.74 18.39 23.23 25.71 18.74 15.09 1.22 26.61]
[100, 011]  [18.39 23.23 25.71 18.74 15.09 1.22 26.61 28.89]
[011, 100]  [23.23 25.71 18.74 15.09 1.22 26.61 28.89 27.62]
[010, 101]  [25.71 18.74 15.09 1.22 26.61 28.89 27.62 12.12]
[010, 101]  [18.74 15.09 1.22 26.61 28.89 27.62 12.12 16.77]
[011, 100]  [15.09 1.22 26.61 28.89 27.62 12.12 16.77 24.43]
[100, 011]  [1.22 26.61 28.89 27.62 12.12 16.77 24.43 21.37]
[101, 010]  [26.61 28.89 27.62 12.12 16.77 24.43 21.37 7.03]
[100, 011]  [28.89 27.62 12.12 16.77 24.43 21.37 7.03 19.24]
[101, 010]  [27.62 12.12 16.77 24.43 21.37 7.03 19.24 13.14]
[100, 011]  [12.12 16.77 24.43 21.37 7.03 19.24 13.14 24.91]
[011, 100]  [16.77 24.43 21.37 7.03 19.24 13.14 24.91 21.79]
[100, 011]  [24.43 21.37 7.03 19.24 13.14 24.91 21.79 4.53]
[011, 100]  [21.37 7.03 19.24 13.14 24.91 21.79 4.53 10.12]
[100, 011]  [7.03 19.24 13.14 24.91 21.79 4.53 10.12 12.83]
[110, 001]  [19.24 13.14 24.91 21.79 4.53 10.12 12.83 12.42]
[101, 010]  [13.14 24.91 21.79 4.53 10.12 12.83 12.42 1.97]
[101, 010]  [24.91 21.79 4.53 10.12 12.83 12.42 1.97 5.13]
```

```
[100, 011]  [21.79 4.53 10.12 12.83 12.42 1.97 5.13 20.26]
[011, 100]  [4.53 10.12 12.83 12.42 1.97 5.13 20.26 25.83]
[010, 101]  [10.12 12.83 12.42 1.97 5.13 20.26 25.83 15.19]
[001, 110]  [12.83 12.42 1.97 5.13 20.26 25.83 15.19 26.59]
[010, 101]  [12.42 1.97 5.13 20.26 25.83 15.19 26.59 7.77]
[011, 100]  [1.97 5.13 20.26 25.83 15.19 26.59 7.77 17.96]
[100, 011]  [5.13 20.26 25.83 15.19 26.59 7.77 17.96 11.07]
[110, 001]  [20.26 25.83 15.19 26.59 7.77 17.96 11.07 12.83]
[100, 011]  [25.83 15.19 26.59 7.77 17.96 11.07 12.83 27.30]
[100, 011]  [15.19 26.59 7.77 17.96 11.07 12.83 27.30 4.29]
[100, 011]  [26.59 7.77 17.96 11.07 12.83 27.30 4.29 5.84]
[100, 011]  [7.77 17.96 11.07 12.83 27.30 4.29 5.84 5.34]
```

As `ts1.gz` contains a time series with 50 elements, we are going to get 43 subsequences from it. Given the previous output, we are going to assign a name to each one of the subsequences – the name is based on the starting index of the subsequence in the original time series – and associate that name with the SAX representation. So, the first subsequence is going to be named S_0, the second S_1, and so on. The last one is going to be called S_{42}.

We begin by creating the structure presented in *Figure 3.6*. In this structure, we have the root node and its children, which are currently empty. Those children are constructed by creating all SAX representations using the specified number of segments and a cardinality value of 2, which means that each SAX word has a single digit, which can be either 0 or 1. At this point, all these children are terminal nodes. This is just the initial version of the iSAX index, which means that at the end of the day, *there might be children without any subsequences* (empty). However, this representation helps us programmatically. Also, bear in mind that *each child of the root node is the root node of its own binary tree*.

At this point, all subsequences are associated with the *maximum cardinality*, based on the iSAX parameters. However, initially, this *maximum cardinality* is reduced to a cardinality value of 2 according to the rule presented in the *Reducing the cardinality of a SAX representation* subsection of *Chapter 2*. Only after a split do we need to use a cardinality other than 2 – and this only happens for the subsequences that are part of the split. Of course, on big iSAX indexes, there are not many terminal nodes that use a *cardinality value of 2 on all their segments* – it should be clear by now that such terminal nodes are directly connected to the root node. Refer to the following figure:

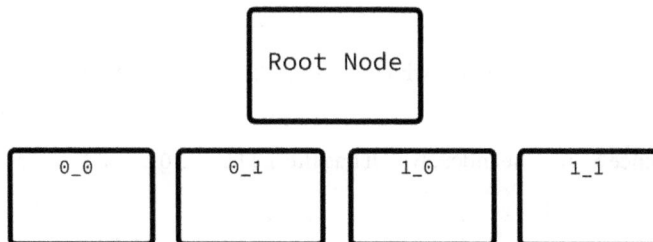

Figure 3.6 – The initial version of iSAX with the root and its children

So, first, we put subsequence S_0 into the index. Before doing that, we need to reduce its cardinality, which goes from [011, 100] to [0, 1]. At this point, we find its matching cardinality in the children of the root based on the reduced cardinality, and we put it there. Now, we put subsequence S_1 into the index, which has the [011, 100] SAX representation, which becomes [0, 1]. Now, we put subsequence S_2 into the index, which has the [100, 011] SAX representation, which becomes [1, 0]. Then, we put subsequence S_3 into the index, which has the [011, 100] SAX representation, which becomes [0, 1].

After that, we have an iSAX index that looks like the one presented in *Figure 3.7*:

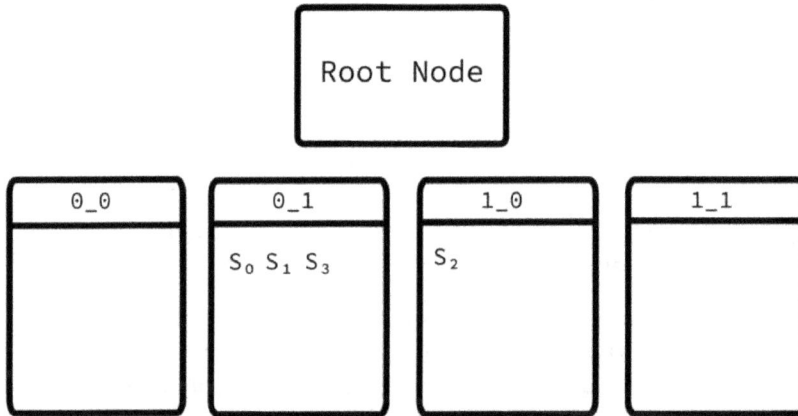

Figure 3.7 – Adding four subsequences to the iSAX index

Now, we put subsequence S_4 into the index, which has the [011, 100] SAX representation, which becomes [0, 1].

Then, we put subsequence S_5 into the index, which has the [100, 011] SAX representation, which becomes [1, 0].

After that, we insert subsequence S_6 into the index, which has the [010, 101] SAX representation, which when reduced becomes [0, 1].

Now, we put subsequence S_7 into the index, which has the [011, 100] SAX representation, which when reduced is [0, 1].

Now, we put subsequence S_8 into the index, which has the [011, 100] SAX representation, which becomes [0, 1].

Then, we put subsequence S_9 into the index, which has the [010, 101] SAX representation, which becomes [0, 1].

Next, we put subsequence S_{10} into the index, which has the [011, 100] SAX representation, which becomes [0, 1].

Currently, the node with the SAX representation of [0, 1] has 9 subsequences (*Figure 3.8*).

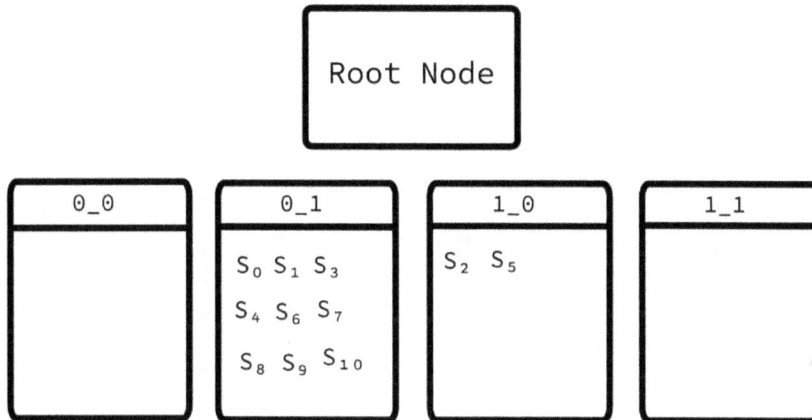

Figure 3.8 – The iSAX index with 11 subsequences

Now, we put subsequence S_{11} into the index, which has the [011, 100] SAX representation, which becomes [0, 1]. Then, we put subsequence S_{12} into the index, which has the [100, 011] SAX representation, which becomes [1, 0].

Now, we insert subsequence S_{13} into the index, which has the [101, 010] SAX representation, which when reduced becomes [1, 0]. After that, we put subsequence S_{14} into the index, which has the [100, 011] SAX representation, which when reduced is [1, 0].

Now, we put subsequence S_{15} into the index, which has the [011, 100] SAX representation, which becomes [0, 1]. Then, we put subsequence S_{16} into the index, which has the [010, 101] SAX representation, which becomes [0, 1].

Now, we put subsequence S_{17} into the index, which has the [010, 101] SAX representation, which becomes [0, 1]. After that, we put subsequence S_{18} into the index, which has the [011, 100] SAX representation, which becomes [0, 1].

Now, we put subsequence S_{19} into the iSAX index, which has the [100, 011] SAX representation, which when reduced is [1, 0]. Then, we put subsequence S_{20} into the iSAX index, which has the [101, 010] SAX representation, which becomes [1, 0].

Now, we put subsequence S_{21} into the iSAX index, which has the [100, 011] SAX representation, which becomes [1, 0].

Then, we put subsequence S_{22} into the iSAX index, which has the [101, 010] SAX representation, which becomes [1, 0].

Now, we put subsequence S_{23} into the iSAX index, which has the [100, 011] SAX representation, which becomes [1, 0].

After that, we put subsequence S_{24} into the iSAX index, which has the [011, 100] SAX representation, which becomes [0, 1]. At this point, the iSAX node with the SAX representation of [0, 1] is full.

Next, we put subsequence S_{25} into the iSAX index, which has the [100, 011] SAX representation, which becomes [1, 0].

Figure 3.9 shows the current version of the iSAX index (the root node is omitted for brevity):

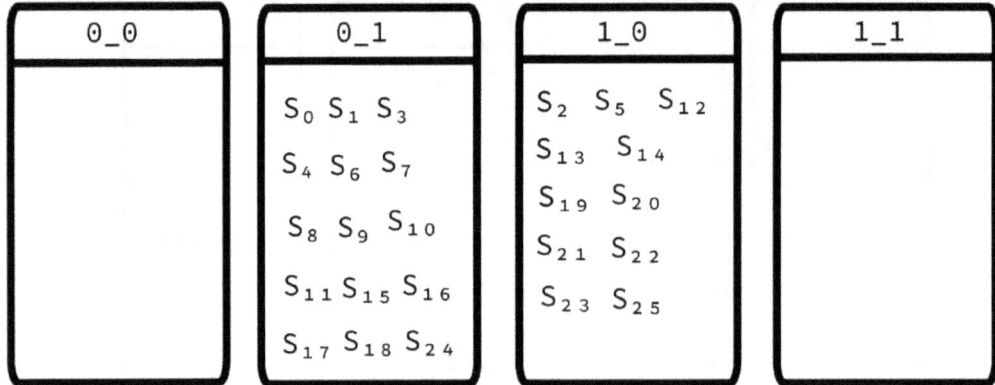

0_0	0_1	1_0	1_1
	S_0 S_1 S_3 S_4 S_6 S_7 S_8 S_9 S_{10} S_{11} S_{15} S_{16} S_{17} S_{18} S_{24}	S_2 S_5 S_{12} S_{13} S_{14} S_{19} S_{20} S_{21} S_{22} S_{23} S_{25}	

Figure 3.9 – Adding subsequences to the iSAX index

Now, we try to insert subsequence S_{26} into the iSAX index, which has the [011, 100] SAX representation, which becomes [0, 1]. At this point, we must perform a split of the [0, 1] terminal node. Therefore, [0, 1] becomes an inner node and two new terminal nodes are created, which become the children of [0, 1]: [00, 1] and [01, 1].

Now, we must calculate the SAX representations of all the existing subsequences of [0, 1] according to the new cardinality in order to put them into one of the two newly created terminal nodes.

The new SAX representations of the subsequences of the previous [0, 1] terminal node are as follows: S_0 --> [01, 1], S_1 --> [01, 1], S_3 --> [01, 1], S_4 --> [01, 1], S_6 --> [01, 1], S_7 --> [01, 1], S_8 --> [01, 1], S_9 --> [01, 1], S_{10} --> [01, 1], S_{11} --> [01, 1], S_{15} --> [01, 1], S_{16} --> [01, 1], S_{17} --> [01, 1], S_{18} --> [01, 1], S_{24} --> [01, 1], and S_{26} --> [01, 1]. If you look closely, you are going to discover that the first SAX word in all previous subsequences is the same: 01. This means that the split is not going to work and we are going to need to perform an additional split. For no particular reason, we are going to keep promoting the first SAX word. Therefore, [01, 1] is going to become an inner node and create two new terminal nodes: [010, 1] and [011, 1]. There is no need to promote the other terminal node ([00, 1]) as there is no issue there.

So, here are the new SAX representations of the previous subsequences: S_0 --> [011, 1], S_1 --> [011, 1], S_3 --> [011, 1], S_4 --> [011, 1], S_6 --> [010, 1], S_7 --> [011, 1], S_8 --> [011, 1], S_9 --> [010, 1], S_{10} --> [011, 1], S_{11} --> [011, 1], S_{15} --> [011, 1], S_{16} --> [010, 1], S_{17} --> [010, 1], S_{18} --> [011, 1], S_{24} --> [011, 1], and S_{26} --> [011, 1].

This split resolves the issue, so the subsequences are put into the appropriate terminal node. *Figure 3.10* shows the latest version of the iSAX index (the root node is omitted for brevity).

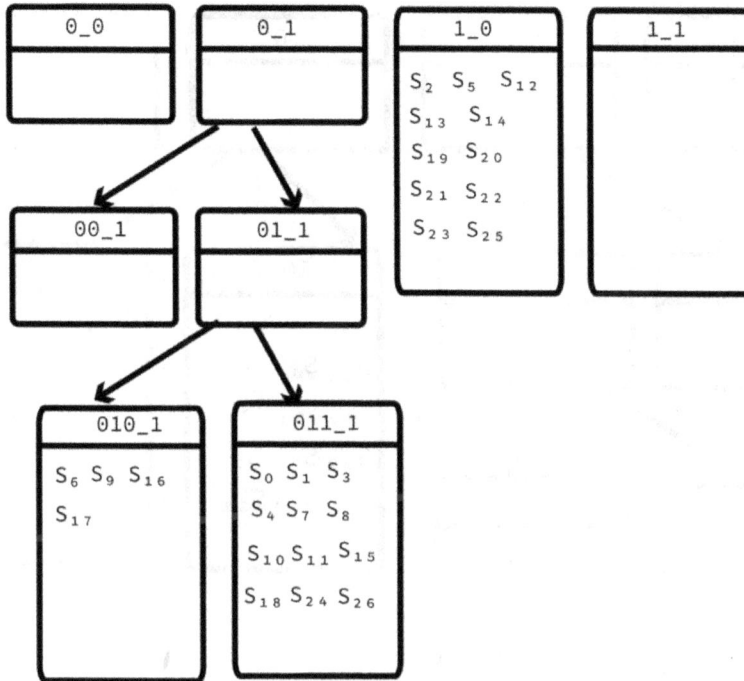

Figure 3.10 – Node splitting of an iSAX index

As stated before, the reason for having to perform a split is that we do not want the number of subsequences stored in a terminal node to be greater than the threshold value.

Now, we put subsequence S_{27} into the index, which has the [100, 011] SAX representation, which becomes [1, 0]. Next, we insert subsequence S_{28} into the index, which has the [110, 001] SAX representation, which becomes [1, 0].

Now, we put subsequence S_{29} into the index, which has the [101, 010] SAX representation, which becomes [1, 0]. Next, we insert subsequence S_{30} into the index, which has the [101, 010] SAX representation, which becomes [1, 0].

Now, we put subsequence S_{31} into the index, which has the [100, 011] SAX representation, which becomes [1, 0]. At this point, we must split the [1, 0] terminal node, which becomes an inner node with two new terminal nodes as its children: [10, 0] and [11, 0]. Therefore, the new SAX representations of the subsequences that were stored in [1, 0] are as follows: S_2 --> [10, 0], S_5 --> [10, 0], S_{12} --> [10, 0], S_{13} --> [10, 0], S_{14} --> [10, 0], S_{19} --> [10, 0], S_{20} --> [10, 0], S_{21} --> [10, 0], S_{22} --> [10, 0], S_{23} --> [10, 0], S_{25} --> [10, 0], S_{27} --> [10, 0], S_{28} --> [11, 0], S_{29} --> [10, 0], S_{30} --> [10, 0], and S_{31} --> [10, 0].

Figure 3.11 shows the latest version of the iSAX index (the root node is omitted for brevity). Notice that [10, 0] terminal node is full and cannot store any more subsequences.

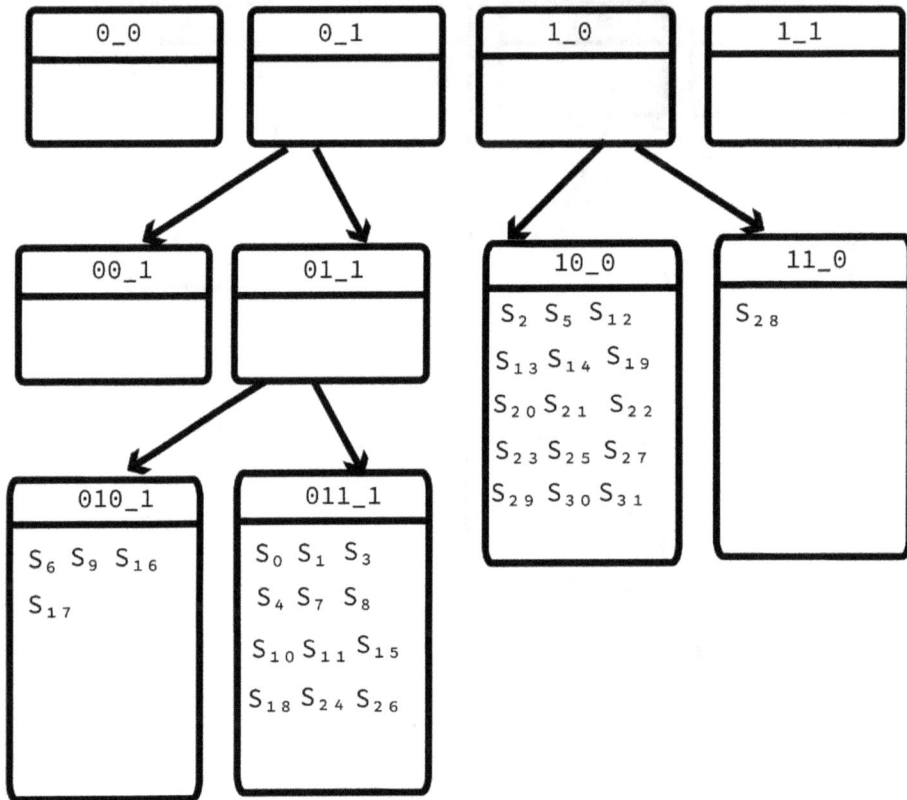

Figure 3.11 – More node splitting in the iSAX index

Now, we put subsequence S_{32} into the index, which has the [011, 100] SAX representation, which becomes [011, 1] (we reduce the cardinality according to our needs). Next, we insert subsequence S_{33} into the index, which has the [010, 101] SAX representation, which when reduced becomes [010, 1].

After that, we put subsequence S_{34} into the index, which has the [001, 110] SAX representation, which becomes [00, 1]. Next, we insert subsequence S_{35} into the index, which has the [010, 101] SAX representation, which becomes [010, 1].

After that, we put subsequence S_{36} into the index, which has the [011, 100] SAX representation, which becomes [011, 1]. Next, we insert subsequence S_{37} into the index, which has the [100, 011] SAX representation, which when reduced becomes [10, 0]. The last subsequence causes a split to the [10, 0] node, which becomes an inner node with two children: [100, 0] and [101, 0].

Therefore, the new SAX representations of the subsequences that were stored in [10, 0] are as follows: S_2 --> [100, 0], S_5 --> [100, 0], S_{12} --> [100, 0], S_{13} --> [101, 0], S_{14} --> [100, 0], S_{19} --> [100, 0], S_{20} --> [101, 0], S_{21} --> [100, 0], S_{22} --> [101, 0], S_{23} --> [100, 0], S_{25} --> [100, 0], S_{27} --> [100, 0], S_{29} --> [101, 0], S_{30} --> [101, 0], S_{31} --> [100, 0], and S_{37} --> [100, 0].

Figure 3.12 shows the updated iSAX index (the root node is omitted for brevity).

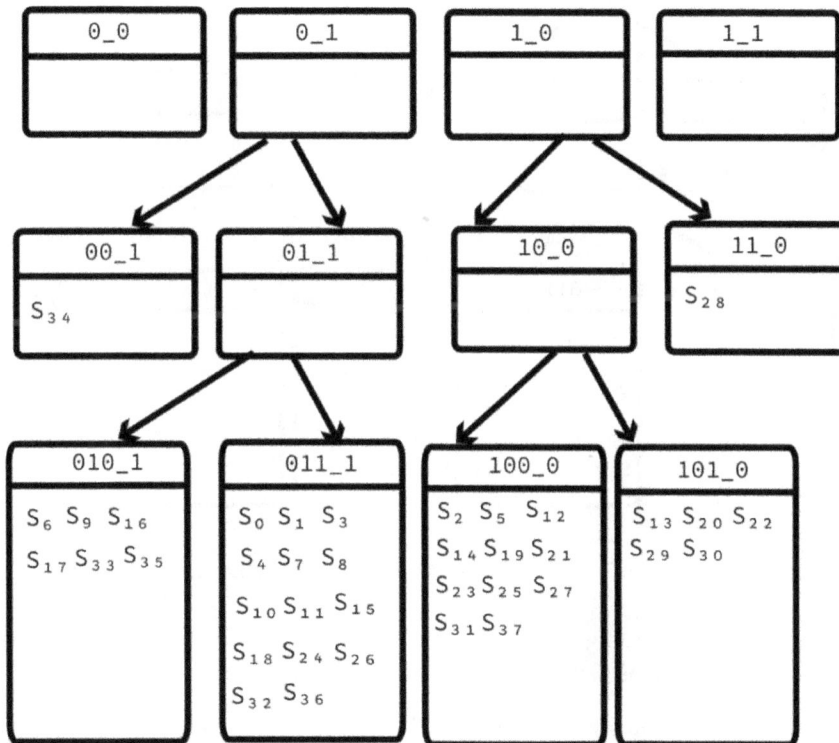

Figure 3.12 – Updated version of the iSAX index with 38 subsequences

Next, we put S_{38} into the index, which has the [110, 001] SAX representation, which becomes [11, 0]. Then, we insert subsequence S_{39} into the index, which has the [100, 011] SAX representation, which becomes [100, 0].

Next, we put subsequence S_{40} into the index, which has the [100, 011] SAX representation, which becomes [100, 0]. After that, we insert subsequence S_{41}, which becomes [100, 0].

Figure 3.13 shows the final version of the iSAX index. If you count the subsequences in the terminal nodes, you are going to find that they are 43, which is the correct number based on the sliding window size and the time series length.

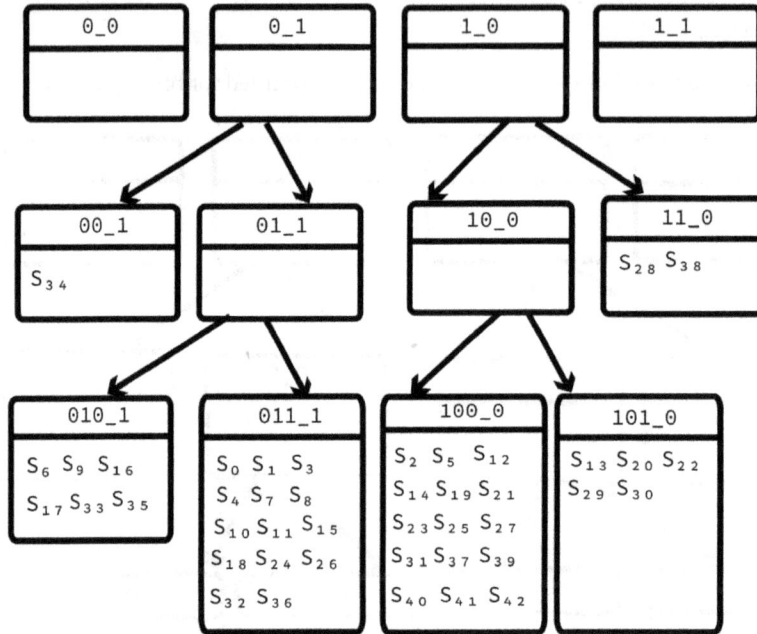

Figure 3.13 – The final version of the iSAX index

You should agree by now that creating an iSAX index manually is a tedious process and nobody should have to do that. This makes it a perfect candidate to be carried out using a computer.

Now that we know how to manually populate an iSAX index, it is time to update the `counting.py` utility from *Chapter 2*.

Updating the counting.py utility

Remember the `counting.py` utility from *Chapter 2*? In this section, we are going to update it and use it for some important tasks. We are not going to completely change the existing functionality or throw away all the existing code. We are going to build on the existing code of the `counting.py` utility, which is a great and productive way to develop new software.

The updated version of the utility can be used for the following tasks:

- Seeing whether a time series can fit into an iSAX index. This computation is based on the existing functionality of `counting.py` combined with a test of whether the values of all dictionary entries are smaller than the threshold value.

- Seeing whether a time series can fit into an iSAX index using more segments or by increasing the threshold. Again, this computation is based on the existing functionality of counting. py, which is enhanced with some extra computations and testing.

- Seeing whether an iSAX index is relatively balanced or not. This can have a great impact on the performance of the index. However, as you already know, indexes are not balanced in general. The idea behind this functionality is based on counting the number of subsequences that go under each child of the root. Put simply, we count the subsequences on nodes that have a cardinality value of 2 on all their SAX words because these are the children of the root, and we print the results on screen. Although this test is not conclusive, it gives us a good idea of how the subsequences are going to be distributed in the iSAX index. Keep in mind that if you use a segments value larger than 4, you are going to get a large amount of output from the utility.

The new version of the counting.py utility, which is called countingv2.py and has been refactored to include a function, contains the following code, which is presented in four parts. The first part is as follows:

```
#!/usr/bin/env python3

import sys
import pandas as pd
from sax import sax

def calculate(ts_numpy, sliding, segments, cardinality):
    KEYS = {}

    length = len(ts_numpy)
    for i in range(length - sliding + 1):
        t1_temp = ts_numpy[i:i+sliding]
        tempSAXword = sax.createPAA(t1_temp, cardinality, segments)
        tempSAXword = tempSAXword[:-1]

        if KEYS.get(tempSAXword) == None:
            KEYS[tempSAXword] = 1
        else:
            KEYS[tempSAXword] = KEYS[tempSAXword] + 1

    return KEYS
```

The calculate() function, which computes the number of subsequences per SAX representation, is called multiple times and is what might slow down the script. Therefore, before seeing countingv2. py in action, let me warn you that the utility might become slow when processing time series with millions of elements.

The second part is as follows:

```python
def main():
    if len(sys.argv) != 6:
        print("TS1 sliding_window cardinality segments threshold")
        sys.exit()

    file = sys.argv[1]
    sliding = int(sys.argv[2])
    cardinality = int(sys.argv[3])
    segments = int(sys.argv[4])
    threshold = int(sys.argv[5])

    if sliding % segments != 0:
        print("sliding MODULO segments != 0...")
        sys.exit()

    if sliding <= 0:
        print("Sliding value is not allowed:", sliding)
        sys.exit()

    if cardinality <= 0:
        print("Cardinality Value is not allowed:",
            cardinality)
        sys.exit()

    ts = pd.read_csv(file, names=['values'],
        compression='gzip')
    ts_numpy = ts.to_numpy()

    # See if it fits
    overflow = False
    KEYS = calculate(ts_numpy, sliding, segments,
        cardinality)
    maxVal = max(KEYS.values())
    if maxVal > threshold:
        overflow = True
```

The condition for an overflow is as follows: *if any node, which is identified by its SAX representation, has more subsequences than the threshold value, then we have an overflow.* Instead of checking all nodes, we get the maximum value of subsequences found in a node (max(KEYS.values())) and we compare that to the given threshold value.

The third part contains the following Python code:

```
# See if we can make it fit or reduce the parameters
if overflow:
    i = 2
    while overflow:
        # We cannot have more segments than the window
        if segments * i > sliding:
            break
        print("Increasing segments to", i * segments)
        overflow = False
        KEYS = calculate(ts_numpy, sliding,
            segments * i, cardinality)
        maxVal = max(KEYS.values())
        if maxVal > threshold:
            overflow = True
            print("Overflow")
            i = 2 * i
        if overflow == False:
            print("New segments:", i * segments)
else:
    print("Threshold can be", max(KEYS.values()))
    print("Reducing cardinality to", cardinality//2)
    overflow = False
    KEYS = calculate(ts_numpy, sliding, segments,
        cardinality//2)
    maxVal = max(KEYS.values())
    if maxVal > threshold:
        print("Cannot reduce cardinality")
    elif overflow == False:
        print("New cardinality:", cardinality//2)
```

The previous code works as follows.

If we have an overflow, then the code doubles the value of the parameter that holds the number of segments while having in mind that the segments parameter cannot be greater than the sliding window size. This is going to keep happening as long as there is an overflow and the segments are not bigger than the sliding window size. If we do not have an overflow in the first place, the code tries to reduce the value of the cardinality parameter by half and see what happens.

The last part is the following:

```
# Now let us see whether the iSAX index is going to be
# balanced or not using a cardinality value of 2
KEYS = calculate(ts_numpy, sliding, segments, 2)
```

```
        minVal = min(KEYS.values())
        maxVal = max(KEYS.values())
        print("Min:", minVal, "Max:", maxVal)
        for k in KEYS.keys():
            print(k, ":", KEYS[k])
```

This last part of the code uses **a cardinality value of 2** to find out under which child of the root each subsequence is going to be placed. In this case, we do not care about overflows.

Let me now present you with some actual uses of countingv2.py. Imagine having a time series with 450,000 elements and you want to know whether it can fit into an iSAX index with 4 segments, a cardinality of 32, and a threshold value of 1500 when using a sliding window of 1024. If you want to perform this test, please create a time series with synthetic data using the synthetic_data.py script from *Chapter 1*. The output of countingv2.py is going to be as follows (your output is going to vary as we are not using the same time series):

```
$ ./countingv2.py 450k.txt.gz 1024 32 4 1500
Threshold can be 317
Reducing cardinality to 16
Cannot reduce cardinality
Min: 11942 Max: 76534
0_1_1_1 : 26080
0_0_1_1 : 57549
0_0_1_0 : 11942
0_1_1_0 : 53496
0_1_0_0 : 13154
1_1_0_0 : 76534
1_0_0_0 : 19933
1_0_1_0 : 25430
1_0_1_1 : 20768
1_0_0_1 : 56345
0_1_0_1 : 22142
1_1_1_0 : 31547
1_1_0_1 : 20707
0_0_0_1 : 13850
```

What does the previous output tell us? The first line tells us that we could have used a threshold value of 317 with the original SAX parameters and the entire time series would have fit into the iSAX index – this means that *the time series can fit into the iSAX index*. However, when reducing the cardinality to 16, the iSAX index cannot fit the entire time series using 4 segments and a threshold value of 1500. Finally, the last lines tell us that 76534 subsequences belong to the [1 1 0 0] SAX representation and 11942 subsequences belong to the [0 0 1 0] SAX representation. The number of subsequences under the rest of the children of the root is between 76534 and 11942. Generally speaking, this is not a bad subsequence distribution for an iSAX index, despite the fact that two of the potential children of the root do not appear in the output: [0 0 0 0] and [1 1 1 1].

Now, let us do the same test using the same sliding window length but this time using an iSAX index with 2 segments, a cardinality of 128, and a threshold value of 1500. This time, the output is as follows:

```
$ ./countingv2.py 450k.txt.gz 1024 128 2 1500
Increasing segments to 4
New segments: 4
Min: 207226 Max: 242251
0_1 : 207226
1_0 : 242251
```

What does the previous output tell us? It tells us that the given iSAX parameters are not enough for hosting the provided time series in an iSAX index. However, increasing the number of segments to 4 would do the trick. The last two lines show a part of the problem: all subsequences go to either [0 1] or [1 0], which means that both the [1 1] and [0 0] children of the root node are not being used at all, which might be a bad situation. Apart from the data itself, this also depends on the sliding window size.

> **Normalization and SAX representation**
>
> After running countingv2.py many times, it came to my attention that it is very rare to get SAX representations with all zeros or all ones, especially when dealing with a cardinality of 2. The main reason for this phenomenon is **normalization**. Due to normalization, the normalized values of a subsequence cannot fall entirely on the left side or right side of 0 unless we are working with a subsequence that is all zeros. In that case, the normalized version is the same as the original and the mean value is exactly 0. According to our way of treating zero, which means that we must decide whether it going to be included in the left area or right area of the 0 breakpoint, we can easily have a SAX representation with all zeros or all ones. However, having both in the same iSAX index is extremely rare or not possible at all. Keep that in mind when interpreting the results of the presented utility.

Last, let us do the same check using an iSAX index with 4 segments, a cardinality of 64, and a threshold value of 250:

```
$ ./countingv2.py 450k.txt.gz 1024 64 4 250
Threshold can be 105
Reducing cardinality to 32
Cannot reduce cardinality
Min: 11942 Max: 76534
0_1_1_1 : 26080
0_0_1_1 : 57549
0_0_1_0 : 11942
0_1_1_0 : 53496
0_1_0_0 : 13154
1_1_0_0 : 76534
```

```
1_0_0_0 : 19933
1_0_1_0 : 25430
1_0_1_1 : 20768
1_0_0_1 : 56345
0_1_0_1 : 22142
1_1_1_0 : 31547
1_1_0_1 : 20707
0_0_0_1 : 13850
```

The previous output informs us that the selected parameters create an iSAX index that can host all the subsequences of the time series, but we cannot reduce the cardinality to 32. Additionally, as we are using the same number of segments, the results regarding the subsequence distribution are exactly the same as in the first example execution of countingv2.py.

When you have such command-line utilities, it is good to experiment with them as much as you can in order to get a better insight into iSAX and how the parameters affect its shape.

So, in this last section of the chapter, we have improved the counting.py utility to make it even more useful.

Summary

In this chapter, we discussed the theory behind the iSAX index and how the SAX representation is related to the iSAX index. We also learned how to manually create an iSAX index given a time series and the required parameters. Additionally, we developed some handy command-line utilities that support iSAX indexes. We now understand the necessary theory behind iSAX and are ready to apply it.

The next chapter uses this chapter as its foundation in order to develop a Python package for computing iSAX indexes.

Useful links

- The web page of Eamonn Keogh: https://www.cs.ucr.edu/~eamonn/
- For research issues regarding time series, read the research paper titled *Big Sequence Management: A glimpse of the Past, the Present, and the Future* by Themis Palpanas
- Themis Palpanas's home page: https://helios2.mi.parisdescartes.fr/~themisp/
- *OK! So* was used for some of the diagrams in this chapter: https://okso.app/

Exercises

Try to work through the following exercises:

- Manually create an iSAX index for the {1, 2, -2, 2, 0, 1, 3, 4} time series with a sliding window of 4 and a SAX representation with 2 segments, a cardinality of 4, and a threshold value of 2.

- Manually create an iSAX index for the {1, 0, 0, 2, 0, 1, -3, 0} time series with a sliding window of 4 and a SAX representation with 2 segments, a cardinality of 4, and a threshold value of 4.

- Manually create an iSAX index for the {1, -1, -1, 2, 0, 1, -3, 0, 4, 6, 8, 10} time series with a sliding window of 6, a SAX representation with 2 segments, a cardinality value of 8, and a threshold value of 4.

- Manually create an iSAX index for the {1, -1, -1, 2, 0, 1, -3, 0, 4, 6, 8, 10} time series with a sliding window of 4 and a SAX representation with 2 segments, a cardinality of 4, and a threshold value of 4.

- Manually create an iSAX index for the {0, 0, 0, 0, 1, -1, -1, 2, 0, 1, -3, 0, 4, 6, 8, 10, 0, 0} time series with a sliding window of 4 and a SAX representation with 2 segments, a cardinality of 4, and a threshold value of 2.

- Create a sample time series with 2,000,000 elements and check whether it can theoretically fit into an iSAX index with 6 segments, a cardinality of 32, and a threshold value of 500.

4
iSAX – The Implementation

Before continuing with this chapter and starting to write code, make sure that you have a good understanding of the information covered in the previous chapter because this chapter is all about implementing iSAX in Python. As a general principle, if you cannot perform a task manually, you are not going to be able to perform it with the help of a computer – the same principle applies to constructing and using an iSAX index.

While reading this chapter, keep in mind that we are creating an iSAX index that *fits in memory* and does not use any external files to store the subsequences of each terminal node. The original iSAX paper suggested the use of external files to store the subsequences of each terminal node mainly because back then, RAM was limited compared to what is the case today, where we can easily have computers with many CPU cores and more than 64 GB of RAM. As a result, the use of RAM makes the entire process much faster than if we used disk files. However, if you do not have lots of RAM on your system and are working with large time series, you might end up using swap space, which slows down the process.

In this chapter, we are going to cover the following main topics:

- A quick look at the `isax` Python package
- The class for storing subsequences
- The class for iSAX nodes
- The class for entire iSAX indexes
- Explaining the missing parts
- Exploring the remaining files
- Using the iSAX Python package

Technical requirements

The GitHub repository with the code can be found at `https://github.com/PacktPublishing/Time-Series-Indexing`. The code for each chapter is in its own directory. Therefore, the code for this chapter can be found in the `ch04` folder and the `ch04` subfolders.

The first section takes a quick look at the Python package that we have developed for the purposes of this chapter, which strangely enough is called `isax`, before going into more detail.

What about bugs?

We have tried our best to give bug-free code. However, bugs might appear in any program, especially when a program is longer than 100 lines! That is why it is crucial to understand the principles behind the operation and construction of an iSAX index and the SAX representation to be able to understand that there might be a small or bigger issue with the code, or be able to port the existing implementation to a different programming language. I wrote the Python version of iSAX using a Java implementation from a colleague as my starting point.

A quick look at the iSAX Python package

In this section, we will take a first look at the iSAX Python package to get a better idea of the supported functionality. Although we will begin with the code from the `sax` package we developed in *Chapter 2*, we are going to rename that package `isax` and create additional source code, which is named `isax.py`.

The structure of the `isax` directory with the Python files is going to be as follows:

```
$ tree isax/
isax/
├── SAXalphabet
├── __init__.py
├── isax.py
├── sax.py
├── tools.py
└── variables.py

1 directory, 6 files
```

So, in total, we have six files. You already know five of them from the `sax` package. The only new one is the `isax.py` source code file, which is the core file for this chapter. Additionally, we are going to add more global variables to the `variables.py` file and some functions to `tools.py`.

The list of methods found in the isax.py file, which excludes __init__() functions, is the following:

```
$ grep -w def isax/isax.py | grep -v __init__
    def insert(self, ts, ISAX):
    def nTimeSeries(self):
    def insert(self, ts_node):
```

The reason that we are talking about methods and not functions is that each function is attached to a Python class, which automatically makes it a method of that class.

Additionally, if we also include __init__() functions in the output, then we might get a good prediction of the number of classes found in that Python file. In that case, you might want to run grep -w def -n1 isax/isax.py instead:

```
$ grep -w def -n1 isax/isax.py
5-class TS:
6:    def __init__(self, ts, segments):
7-        self.index = 0
--
11-class Node:
12:    def __init__(self, sax_word):
13-        self.left = None
--
19-    # Follow algorithm from iSAX paper
20:    def insert(self, ts, ISAX):
21-        # Accessing a subsequence
--
127-
128:    def nTimeSeries(self):
129-        if self.terminalNode == False:
--
141-class iSAX:
142:    def __init__(self):
143-        # This is now a hash table
--
148-
149:    def insert(self, ts_node):
150-        # Array with number of segments
```

So, we have three classes, named TS, Node, and iSAX.

The next sections are going to discuss the methods of isax.py in relation to the class that they belong to.

The class for storing subsequences

In this subsection, we are going to explain the Python class used for **storing subsequences**, which is named TS. The definition of the class is as follows:

```
class TS:
    def __init__(self, ts, segments):
        self.index = 0
        self.ts = ts
        self.maxCard = sax.createPAA(ts,
            variables.maximumCardinality, segments)
```

When defining an object of that class, we need to provide the ts parameter, which is a subsequence stored as a NumPy array, and the number of segments using the segments parameter. After that, the maxCard field is automatically initialized with the SAX representation of that subsequence with the maximum cardinality. The index parameter is optional and keeps the place of the subsequence in the original time series. iSAX does not use the index parameter but it is good to have such a field.

This class not have any methods attached to it, which is not the case with the Node class that is presented next.

The class for ISAX nodes

In this subsection, we will explain the Python structure used for *keeping both inner and terminal nodes*. This is an important part of the package and its functionality:

```
class Node:
    def __init__(self, sax_word):
        self.left = None
        self.right = None
        self.terminalNode = False
        self.word = sax_word
        self.children = [TS] * variables.threshold
```

If we are dealing with an inner node, then the terminalNode field is set to False. However, if the Boolean value of the terminalNode field is set to True, then we are dealing with a terminal node.

The word field holds the SAX representation of the node. Lastly, the left and right fields are links to the two children of an inner node, whereas the children field is a list that holds the subsequences of a terminal node.

The Node class has two methods:

- insert(): This method is used for adding subsequences to a node

- nTimeSeries(): This method is used for counting the number of subsequences stored in a terminal node

Next, let us talk about the class for representing entire iSAX indexes.

The class for entire iSAX indexes

This last class of the `isax` package is used to represent entire iSAX indexes:

```
class iSAX:
    def __init__(self):
        # This is now a hash table
        self.children = {}
        # HashTable for storing Nodes
        self.ht = {}
        self.length = 0
```

The `children` field holds the children of the root node – in reality, the instances of the iSAX class are roots of iSAX indexes.

The `ht` field, which is a dictionary, holds all the nodes of the iSAX index. Each key is a SAX representation of a node, which is *unique*, and each value is a `Node` instance. Lastly, the `length` field holds the number of subsequences stored in the iSAX index and is an optional field.

The `iSAX` class has only one method, which is called `insert()` and is used to insert subsequences into the iSAX index.

> **Why are we using these three classes?**
>
> The implementation of the iSAX index contains three different entities: subsequences, nodes, and the iSAX index itself, which is represented by the root node of the index. iSAX contains nodes and nodes contain other nodes or subsequences. Each of these entities has its own class.

So far, we know the details of the Python classes used in our package. The next section is about implementing the missing parts.

Explaining the missing parts

In this section, we are going to show the implementations of the class methods. We begin with the `insert()` function of the `iSAX` class, which should not be confused with the `insert()` function of the `Node` class. In Python and many other programming languages, classes are independent entities, which means they can have methods with the same name as long as they are unique inside the class namespace.

We are going to present the code of `Node.insert()` in eight parts. The method accepts two parameters – apart from `self`, which denotes the current `Node` object – which are the subsequences we are trying to insert and the iSAX index that the `Node` instance belongs to.

Why do we need an iSAX instance as a parameter? We need that in order to be able to add new nodes to the iSAX index by accessing iSAX.ht.

The first part of insert() is the following:

```
# Follow algorithm from iSAX paper
def insert(self, ts, ISAX):
    # Accessing a subsequence
    variables.nSubsequences += 1
    if self.terminalNode:
        if self.nTimeSeries() == variables.threshold:
            variables.nSplits += 1

            # Going to duplicate self Node
            temp = Node(self.word)
            temp.children = self.children
            temp.terminalNode = True

            # The current Terminal node becomes
            # an inner node
            self.terminalNode = False
            self.children = None
```

The first thing that insert() does is check whether we are working with a terminal node or not. This happens because if we are dealing with a terminal node, we are going to try to store the given subsequence in the terminal node without any other delay. The second check is whether the terminal node is full or not. If it is full, then *we have a split*. First, we duplicate the current node with the temp = Node(self.word) statement and the current terminal node becomes an inner node by changing the value of terminalNode to False. At this point, we have to create two new empty nodes, which are going to become the two children of the current node – the former takes place in the code excerpt that follows.

The second part of insert() is as follows:

```
# Create TWO new Terminal nodes
new1 = Node(temp.word)
new1.terminalNode = True
new2 = Node(temp.word)
new2.terminalNode = True

n1Segs = new1.word.split('_')
n2Segs = new2.word.split('_')
```

In the previous code, we create two new terminal nodes, which are going to be the children of the node that is going to be split. Both of these new nodes currently have the same SAX representation as the node that is going to be split and become an inner node. The change to their SAX representations, which signifies the split, is going to take place in the code that follows.

The third part contains the following code:

```
# This is where the promotion
# strategy is selected
if variables.defaultPromotion:
    tools.round_robin_promotion(n1Segs)
else:
    tools.shorter_first_promotion(n1Segs)

# New SAX_WORD 1
n1Segs[variables.promote] =
    n1Segs[variables.promote] + "0"
# CONVERT it to string
new1.word = "_".join(n1Segs)

# New SAX_WORD 2
n2Segs[variables.promote] =
    n2Segs[variables.promote] + "1"
# CONVERT it to string
new2.word = "_".join(n2Segs)

# The inner node has the same
# SAX word as before but this is
# not true for the two
# NEW Terminal nodes, which should
# be added to the Hash Table
ISAX.ht[new1.word] = new1
ISAX.ht[new2.word] = new2

# Associate the 2 new Nodes with the
# Node that is being splitted
self.left = new1
self.right = new2
```

At the beginning of the code excerpt, we deal with the promotion strategy, which is implemented in the tools.py file, which is explained in *The tools.py file* section and has to do with defining the SAX word (segment) that is going to be promoted.

After that, the code creates the two SAX representations of a split using two string operations – this is the main reason that we store SAX words as strings and that we are using a list to hold an entire SAX presentation. After that, we convert the SAX representations into strings stored in new1.word and new2.word and then put the respective nodes into the iSAX index using ISAX.ht[new1.word] = new1 and ISAX.ht[new2.word] = new2. The keys for finding these two nodes in the iSAX.ht Python dictionary are their own SAX representations. The last two statements of the code associate the two new terminal nodes with the inner node by defining the left and right fields of the inner node and, therefore, signifying its two children.

The fourth code part of the Node.insert() method is as follows:

```
# Check all TS in original node
# and put them
# in one of the two children
#
# This is where the actual
# SPLITTING takes place
#
for i in range(variables.threshold):
    # Accessing a subsequence
    variables.nSubsequences += 1

    # Decrease TS.maxCard to
    # current Cardinality
    tempCard =
        tools.promote(temp.children[i],
        n1Segs)

    if tempCard == new1.word:
        new1.insert(temp.children[i], ISAX)
    elif tempCard == new2.word:
        new2.insert(temp.children[i], ISAX)
    else:
        if variables.overflow == 0:
            print("OVERFLOW:", tempCard)
        variables.overflow =
            variables.overflow + 1

# Now insert the INITIAL TS node!
# self is now an INNER node
self.insert(ts, ISAX)

if variables.defaultPromotion:
    # Next time, promote the next segment
```

```
            Variables.promote = (variables.promote
                 + 1) % variables.segments
```

We know for sure that after we split the subsequences, which were previously stored in the terminal node that has become an inner node, we are not going to have an overflow. However, we still need to call `self.insert(ts, ISAX)` to insert the subsequence that created the overflow previously and see what is going to happen.

The last `if` checks whether we are using the default promotion strategy, which is the Round Robin strategy, and in that case, it changes the promotion segment to the next in order.

But how do we know whether there is an overflow situation? If after promoting a subsequence to a higher cardinality than its current one (`tempCard`) that subsequence cannot be assigned to any of the two newly created terminal nodes (`new1.word` or `new2.word`), we know that it has not been promoted. Therefore, we have an overflow condition. This is implemented in the `else:` branch of the `if tempCard == new1.word:` block.

The fifth part of `Node.insert()` is next:

```
        else:
            # TS is added if we have a Terminal node
            self.children[self.nTimeSeries()] = ts
```

The code of the previous `else` is executed when we are dealing with a terminal node that is not full. So, we store the given subsequence in the `children` list – this is the ideal way to add a new subsequence to an iSAX index.

The sixth part of the `insert()` function is as follows:

```
        else:
            # Otherwise, we are dealing with an INNER node
            # and we should add it to the
            # INNER node by trying
            # to find an existing terminal node
            # or create a new one
            # See whether it is going to be
            # included in the left
            # or the right child
            left = self.left
            right = self.right
```

If we are working with an inner node, we have to decide whether the subsequence is going to go to the left or right child according to its SAX representation in order to finally find the terminal node that is going to store that subsequence. This is where the process begins.

The seventh part contains the following code:

```
leftSegs = left.word.split('_')
# Promote
tempCard = tools.promote(ts, leftSegs)
```

In the previous code, we change (decrease) the maximum cardinality of the subsequence to fit the cardinality of the left node – we could have used the right node as both nodes use the same cardinality. The tempCard variable that holds that new cardinality is going to be used to decide the path that the subsequence is going to follow in the tree until it finds the appropriate terminal node.

The last part of Node.insert() is as follows:

```
if tempCard == left.word:
    left.insert(ts, ISAX)
elif tempCard == right.word:
    right.insert(ts, ISAX)
else:
    if variables.overflow == 0:
        print("OVERFLOW:", tempCard, left.word,
            right.word)
    variables.overflow = variables.overflow + 1


return
```

If tempCard does not match the SAX representation of the left or right node, then we know that *it has not been promoted*, which means that we have an overflow condition.

This is the logic behind the implementation of Node.insert() – there exist many comments in the code that you can read, and you can add your own print() statements to understand the flow even better.

> **Why are we storing the maximum cardinality in its subsequence?**
>
> The reason for storing the maximum cardinality of this subsequence is that we can easily decrease that maximum cardinality without having to make difficult calculations such as computing a new SAX representation from scratch. This small optimization makes splitting operations much faster.

The other method of the Node class is called nTimeSeries() and has the following implementation:

```
def nTimeSeries(self):
    if self.terminalNode == False:
        print("Not a terminal node!")
        return
```

```
        n = 0
        for i in range(0, variables.threshold):
            if type(self.children[n]) == TS:
                n = n + 1


        return n
```

The presented function returns the number of subsequences stored in a terminal node. First, nTimeSeries() makes sure that we are dealing with a terminal node before iterating over the contents of the children list. If the data type of the stored value is TS, then we have a subsequence.

After that, we are going to discuss and explain the insert() method of the iSAX class, which is presented in three parts. The iSAX.insert() method is the method that is being called when we want to add a subsequence to an iSAX index.

The first part of iSAX.insert() is as follows:

```
    def insert(self, ts_node):
        # Array with number of segments
        # For cardinality 1
        segs = [1] * variables.segments


        # Get cardinality 1 from ts_node
        # in order to find its main subtree
        lower_cardinality = tools.lowerCardinality(segs,
            ts_node)


        lower_cardinality_str = ""
        for i in lower_cardinality:
            lower_cardinality_str = lower_cardinality_str +
                "_" + i


        # Remove _ at the beginning
        lower_cardinality_str = lower_cardinality_str[
            1:len(lower_cardinality_str)]
```

This first part of the code finds the child of the root where the given subsequence is going to be placed. The lower_cardinality_str value is used as the key for finding the relevant child of the root node – the tools.lowerCardinality() function is explained in a bit.

The second part of iSAX.insert() contains the following code:

```
        # Check whether the SAX word with CARDINALITY 1
        # exists in the Hash Table.
        # If not, create it and update Hash Table
        if self.ht.get(lower_cardinality_str) == None:
```

```
                    n = Node(lower_cardinality_str)
                    n.terminalNode = True
                    # Add it to the hash table
                    self.children[lower_cardinality_str] = n
                    self.ht[lower_cardinality_str] = n
                    n.insert(ts_node, self)
```

If the child of the root with the `lower_cardinality_str` SAX representation cannot be found, we create the respective root child and add it to the `self.children` hash table (dictionary) and call `insert()` to put the given subsequence there.

The last part of `iSAX.insert()` is the following:

```
            else:
                n = self.ht.get(lower_cardinality_str)
                n.insert(ts_node, self)

        return
```

If the child of the root with the `lower_cardinality_str` SAX representation exists, then we try to insert that subsequence, thereby calling `insert()`.

At this point, we go from the `iSAX` class level to the `Node` class level.

But `isax.py` is not the only file with new code. The next section shows the additions and changes to the remaining package files that complete the implementation.

Exploring the remaining files

Apart from the `isax.py` file, the `isax` Python package is constructed of more source code files, mainly because it is based on the `sax` package. We will begin with the `tools.py` file.

The tools.py file

There are some additions to the `tools.py` source code file compared to the version we first saw in *Chapter 2, which* mainly have to do with the promotion strategy. As said before, we support two promotion strategies: Round Robin and from left to right.

The Round Robin strategy is implemented here:

```
def round_robin_promotion(nSegs):
    # Check if there is a promotion overflow
    n = power_of_two(variables.maximumCardinality)
    t = 0

    while len(nSegs[variables.promote]) == n:
```

```
    # Go to the next SAX word and promote it
    Variables.promote = (variables.promote + 1) %
        variables.segments
    t += 1
    if t == variables.segments:
        if variables.overflow == 0:
            print("Non recoverable Promotion overflow!")
        return
```

In the Round Robin case, we try to find the first segment on the right of the segment that was used in the previous promotion with fewer digits than the digits specified by the maximum cardinality (a segment that is not full). If the previous promotion took place in the last segment, then we go back to the first segment and begin from scratch. In order to compute the number of binary digits of the maximum cardinality (the length of the SAX word), we use the power_of_two() function, which returns 3 for a cardinality of 8, 4 for a cardinality of 16, and so on. If we iterate over all the segments of the given SAX representation (nSegs) and all have the maximum length, we know that we have an overflow condition.

The left-to-right strategy, which is also called **shorter first**, is implemented here:

```
def shorter_first_promotion(nSegs):
    length = len(nSegs)
    pos = 0
    min = len(nSegs[pos])
    for i in range(1,length):
        if min > len(nSegs[i]):
            min = len(nSegs[i])
            pos = i

    variables.promote = pos
```

The left-to-right promotion strategy iterates over all the segments of the given SAX representation variable (nSegs) starting from the left and going to the right and finds the leftmost one with the minimum length. As a result, if both the second and third segments have the same minimum length, the strategy is going to select the second one because it is the leftmost available. After that, it sets variables.promote to the selected segment value.

Next, we are going to talk about two additional functions that reside in tools.py, which are called promote() and lowerCardinality().

The promote() function is implemented as follows:

```
def promote(node, segments):
    new_sax_word = ""
    max_array = node.maxCard.split("_")[
        0:variables.segments]
```

```
# segments is an array
#
for i in range(variables.segments):
    t = len(segments[i])
    new_sax_word = new_sax_word + "_" +
        max_array[i][0:t]

# Remove _ from the beginning of the new_sax_word
new_sax_word = new_sax_word[1:len(new_sax_word)]
return new_sax_word
```

The `promote()` function copies the length of the digits of the segments of an existing SAX representation (node) to a given subsequence (s) in order for both to have the same cardinalities in all their SAX words. This *allows us to compare* these two SAX representations.

The implementation of `lowerCardinality()` is as follows:

```
def lowerCardinality(segs, ts_node):
    # Get Maximum Cardinality
    max = ts_node.maxCard
    lowerCardinality = [""] * variables.segments

    # Because max is a string, we need to split.
    # The max string has an
    # underscore character at the end.
    max_array = max.split("_")[0:variables.segments]

    for i in range(variables.segments):
        t = segs[i]
        lowerCardinality[i] = max_array[i][0:t]

    return lowerCardinality
```

The `lowerCardinality()` function lowers the cardinalities of a node *in all of its SAX words* (segments). This is mainly needed by `iSAX.insert()` to put a subsequence into the appropriate child of the root. After we put a subsequence into the appropriate child of the root, we promote a single segment of the SAX representation of the subsequence at a time to find out its place in the iSAX index. Remember that the keys to all iSAX nodes are SAX representations that usually have different cardinalities in their segments.

> **How to test individual functions**
>
> Personally, I prefer to create small command line utilities to test complex functions on their own, understand their operation, and maybe discover bugs!

Let us make two small command-line utilities to showcase the use of `promote()` and `lowerCardinality()` in more detail.

First, we demonstrate the `promote()` function in the `usePromote.py` utility, which contains the following code:

```
#!/usr/bin/env python3

from isax import variables
from isax import isax
import numpy as np

variablesPromote = 0
maximumCardinality = 8
segments = 4

def promote(node, s):
    global segments

    new_sax_word = ""
    max_array = node.maxCard.split("_")[0:segments]

    for i in range(segments):
        t = len(s[i])
        new_sax_word = new_sax_word + "_" +
            max_array[i][0:t]

    new_sax_word = new_sax_word[1:len(new_sax_word)]
    return new_sax_word
```

It is important to remember that the `promote()` function mimics the lengths of the segments of an existing SAX representation by decreasing the maximum SAX representation of a subsequence (s) to match the given SAX representation stored in the node parameter.

The rest of `usePromote.py` is the following:

```
def main():
    global variablesPromote
    global maximumCardinality
    global segments

    variables.maximumCardinality = maximumCardinality
    ts = np.array([1, 2, 3, 4])
    t = isax.TS(ts, segments)

    SAX_WORD = "0_0_1_1_"
```

```
        Segs = SAX_WORD.split('_')
        print("Max cardinality:", t.maxCard)

        SAX_WORD = "00_0_1_1_"
        Segs = SAX_WORD.split('_')
        print("P1:", promote(t, Segs))

        SAX_WORD = "000_0_1_1_"
        Segs = SAX_WORD.split('_')
        print("P2:", promote(t, Segs))

        SAX_WORD = "000_01_1_1_"
        Segs = SAX_WORD.split('_')
        print("P3:", promote(t, Segs))

        SAX_WORD = "000_011_1_100_"
        Segs = SAX_WORD.split('_')
        print("P4:", promote(t, Segs))

if __name__ == '__main__':
    main()
```

Everything is hardcoded in usePromote.py because we just want to know more about the use of promote() and nothing else. However, as promote() has many dependencies in the isax package, we must put its entire implementation in our script and make the necessary changes to the Python code.

Given a subsequence, ts, and a TS class instance, t, we can calculate the SAX representation of ts using the maximum cardinality and then decrease it to match the cardinalities of other SAX words.

Running usePromote.py generates the following output:

```
$ ./usePromote.py
Max cardinality: 000_010_101_111_
P1: 00_0_1_1
P2: 000_0_1_1
P3: 000_01_1_1
P4: 000_010_1_111
```

The output shows that the maximum cardinality (000_010_101_111) of the given subsequence has been decreased to match the cardinalities of four other SAX words.

After that, we demonstrate the `lowerCardinality()` function in the `useLCard.py` utility, which comes with the following code:

```python
#!/usr/bin/env python3

from isax import variables
from isax import tools
from isax import isax
import numpy as np

def main():
    global maximumCardinality
    global segments

    # Used by isax.TS()
    variables.maximumCardinality = 8
    variables.segments = 4
    ts = np.array([1, 2, 3, 4])
    t = isax.TS(ts, variables.segments)

    Segs = [1] * variables.segments
    print(tools.lowerCardinality(Segs ,t))
    Segs = [2] * variables.segments
    print(tools.lowerCardinality(Segs ,t))
    Segs = [3] * variables.segments
    print(tools.lowerCardinality(Segs ,t))

if __name__ == '__main__':
    main()
```

This time, we do not put the implementation of `lowerCardinality()` in our code because it has fewer dependencies and can be used directly from the `tools.py` file. The parameter that we pass to `lowerCardinality()` is *the number of digits* that we want to get in each SAX word. So, 1 means one digit, which means a cardinality of 2 1, and 3 means three digits, which computes to a cardinality of 2 3.

Once again, everything is hardcoded in `useLCard.py` because we just want to know more about the use of `lowerCardinality()` and nothing more. Running `useLCard.py` produces the following output:

```
$ ./useLCard.py
['0', '0', '1', '1']
['00', '01', '10', '11']
['000', '010', '101', '111']
```

So, given a subsequence with a SAX representation of `000_010_101_111`, we calculate its SAX representations for the cardinalities of 2, 4, and 8.

Next, we are going to show the changes to `variables.py`, which is the file that holds global variables that can be accessed by all the files of the package or the utilities that use the `isax` package.

The variables.py file

This subsection presents the contents of the updated `variables.py` file, which contains variables that are accessible from anywhere in the code.

> **How much functionality is enough?**
>
> Keep in mind that sometimes we might need to include functionality that is going to help with debugging or might be needed in the future, and therefore, we might need to include variables or implement functions that are not going to be used right away or all the time. Just remember to keep a good balance between wanting to support everything and please everyone, which is impossible, and wanting to support the absolute minimum functionality, which usually lacks flexibility.

The contents of the `variables.py` file are the following:

```
# This file includes all variables for the isax package
#

maximumCardinality = 32
breakpointsFile = "SAXalphabet"
# Breakpoints in breakpointsFile
elements = ""
slidingWindowSize = 16
segments = 0

# Maximum number of time series in a terminal node
threshold = 100

# Keeps number of splits
nSplits = 0

# Keep number of accesses of subsequences
nSubsequences = 0

# Currently supporting TWO promotion strategies
defaultPromotion = True
```

```
# Number of overflows
overflow = 0

# Floating point precision
precision = 5

# Segment to promote
promote = 0
```

The variables.promote variable defines the SAX word that is going to be promoted next if there is such a need. Put simply, we create the SAX representation of the two nodes of a split based on the value of variables.promote – we promote the segment defined by the value of variables. promote. Every time we have a split, variables.promote is updated according to the promotion (splitting) strategy and gets ready for the next split.

Should you wish to see the changes between two versions of the same file, you can use the diff(1) utility. In our case, the difference between the variables.py file found in the ch03 directory and the current version is the following:

```
2c2
< # This file includes all variables for the sax package
---
> # This file includes all variables for the isax package
13a14,16
> # Breakpoints in breakpointsFile
> elements = ""
>
20,21c23,24
< # Breakpoints in breakpointsFile
< elements = ""
---
> # Maximum number of time series in a terminal node
> threshold = 100
22a26,37
> # Keeps number of splits
> nSplits = 0
>
> # Keeps number of accesses of subsequences
> nSubsequences = 0
>
> # Currently supporting TWO promotion strategies
> defaultPromotion = True
>
> # Number of overflows
```

```
> overflow = 0
>
24a40,42
>
> # Segment to promote
> promote = 0
```

Lines beginning with > show the contents of ch04/isax/variables.py, whereas lines beginning with < show statements from ch03/sax/variables.py.

The next subsection discusses sax.py, which did not change that much.

The sax.py file

The sax.py file does have any practical changes. However, we should make changes to its import statements as it is no longer an autonomous package but a part of another package with a different name. Therefore, we need to change the following two statements:

```
from sax import sax
from sax import variables
```

We replace them with these statements:

```
from isax import sax
from isax import variables
```

Apart from that, there is no need for additional changes.

Now that we know the source code of the isax package, it is time to see that code in action.

Using the iSAX Python package

In this section, we are going to use the isax Python package to develop practical command-line utilities. But first, we are going to learn how to read the iSAX parameters from the users.

Reading the iSAX parameters

This subsection illustrates how to read the iSAX parameters, including the filenames with the time series, and how to give default values to some of them. Although we saw relevant code in *Chapter 2*, this time, the process is explained in more detail. Additionally, the code is also going to show how we use these input parameters to set up the relevant variables located inside the ./isax/variables.py file. As a reminder, variables stored in ./isax/variables.py, or other similar files – it just happens that we are using ./isax/variables.py – are accessible from anywhere in our code as long as we have successfully imported the relevant file.

> **What we need to create an iSAX index**
>
> As a reminder, to create an iSAX index, we need a time series and a threshold value, which is the maximum number of subsequences that a terminal node can hold, as well as a segment value and a cardinality value. Last, we need a sliding window size.

As a rule of thumb, when working with global variables, it is better to use long and descriptive names. Additionally, it is also a good practice to give default values to global parameters.

The Python code of `parameters.py` is shown here:

```python
#!/usr/bin/env python3

import argparse
from isax import variables

def main():
    parser = argparse.ArgumentParser()
    parser.add_argument("-s", "--segments",
        dest = "segments", default = "4",
        help="Number of Segments", type=int)
    parser.add_argument("-c", "--cardinality",
        dest = "cardinality", default = "32",
        help="Cardinality", type=int)
    parser.add_argument("-w", "--window", dest = "window",
        default = "16", help="Sliding Window Size",
        type=int)
    parser.add_argument("TS1")

    args = parser.parse_args()

    variables.segments = args.segments
    variables.maximumCardinality = args.cardinality
    variables.slidingWindowSize = args.window

    windowSize = variables.slidingWindowSize
    maxCardinality = variables.maximumCardinality
    f1 = args.TS1
    print("Time Series:", f1, "Window Size:", windowSize)
    print("Maximum Cardinality:", maxCardinality,
        "Segments:", variables.segments)

if __name__ == '__main__':
    main()
```

All the work is done by the `argparse` package and the `parser.add_argument()` statements that are used for defining command-line parameters and options. The `dest` parameter defines the name of the parameter – this name is going to be used later to read the value of the parameter.

One of the other parameters of `parser.add_argument()` is called `type` and allows us to define the data type of the parameter. This can save you from lots of issues and code for converting strings into actual values, so use `type` when possible.

After that, we call `parser.parse_args()` and we are ready to read any `rgparse` parameter we want.

Running `parameters.py` produces the following output:

```
$ ./parameters.py -s 2 -c 32 -w 16 ts1.gz
Time Series: ts1.gz Window Size: 16
Maximum Cardinality: 32 Segments: 2
```

In case of an error, `parameters.py` generates the following output:

```
$ ./parameters.py -s 1 -c cardinality ts1.gz
usage: parameters.py [-h] [-s SEGMENTS] [-c CARDINALITY] [-w WINDOW]
TS1
parameters.py: error: argument -c/--cardinality: invalid int value:
'cardinality'
```

In this case, the error is that the `cardinality` parameter is a string, whereas we are expecting an integer. The error output is very informative.

If a required parameter is missing, `parameters.py` generates the following output:

```
$ ./parameters.py
usage: parameters.py [-h] [-s SEGMENTS] [-c CARDINALITY] [-w WINDOW]
TS1
parameters.py: error: the following arguments are required: TS1
```

The next section shows how we process the subsequences of a time series in order to create an iSAX index.

How to process subsequences to create an iSAX index

This is a really important subsection because here, we explain the Python structure that is used to store the data for each subsequence of an iSAX index.

> **The code does not lie!**
>
> If you have doubts about the fields and the data stored in each subsequence, check out the Python code to learn more. The documentation might lie but the code never does.

The `subsequences.py` script shows how we create the subsequences, how we store them in a Python data structure, and how we might process them:

```python
#!/usr/bin/env python3

import argparse
import numpy as np
import pandas as pd
from isax import sax
from isax import variables

class TS:
    def __init__(self, ts, index):
        self.ts = ts
        self.sax = sax.createPAA(ts,
            variables.maximumCardinality,
            variables.segments)
        self.index = index

def main():
    parser = argparse.ArgumentParser()
    parser.add_argument("-w", "--window", dest = "window",
        default = "16", help="Sliding Window Size",
        type=int)
    parser.add_argument("-s", "--segments",
        dest = "segments", default = "4",
        help="Number of Segments", type=int)
    parser.add_argument("-c", "--cardinality",
        dest = "cardinality", default = "32",
        help="Cardinality", type=int)
    parser.add_argument("TS")

    args = parser.parse_args()
    windowSize = args.window
    variables.segments = args.segments
    variables.maximumCardinality = args.cardinality
    file = args.TS
```

We define the TS class again and use that version in `subsequences.py` in order to be able to make more changes to the TS class without the danger of altering the code of the `isax` package. Before now, we have read the parameters of the program and we are ready to read the time series:

```python
    ts = pd.read_csv(file, names=['values'],
        compression='gzip', header = None)
```

```
ts_numpy = ts.to_numpy()
length = len(ts_numpy)
```

Currently, we have the time series stored as a NumPy array using the `ts_numpy` variable:

```
# Split sequence into subsequences
n = 0
for i in range(length - windowSize + 1):
    # Get the actual subsequence
    ts = ts_numpy[i:i+windowSize]
    # Create new TS node based on ts
    ts_node = TS(sax.normalize(ts), i)
    n = n + n
```

The `for` loop splits the time series into subsequences based on the sliding window size. The normalized version of each subsequence is stored in a `TS()` structure that has three members: the normalized version of the subsequence (`ts`), the SAX representation of the subsequence (`sax`), and the place of the subsequence in the time series (`index`). The last member of the `TS()` structure allows us to find the original version of the subsequence, should we want to do so.

Now, check the following code:

```
    print("Created", n, "TS() nodes")

if __name__ == '__main__':
    main()
```

After we finish, the script prints the number of subsequences that have been processed.

Only after we have stored the SAX representation of a subsequence in its Python structure based on the maximum cardinality are we ready to put that subsequence into the iSAX index. So, the next step, which is not presented here, is putting each `TS()` node into an iSAX index.

The output of `subsequences.py` gives you information about the number of subsequences that have been processed:

```
$ ./subsequences.py ts1.gz
Created 35 TS() nodes
```

In summary, this is the way we are going to process subsequences in order to add them to an iSAX index. In the next subsection, we are going to create our first iSAX index!

Creating our first iSAX index

In this section, we are going to create an iSAX index for the first time. But first, we are going to present the Python utility for doing that. The Python code of `createiSAX.py` is presented in four parts. The first part is the following:

```python
#!/usr/bin/env python3

from isax import variables
from isax import isax
from isax import tools
from isax import sax

import sys
import pandas as pd
import numpy as np

import time
import argparse

def main():
    parser = argparse.ArgumentParser()

    parser.add_argument("-s", "--segments",
        dest = "segments", default = "16",
        help="Number of Segments", type=int)
    parser.add_argument("-c", "--cardinality",
        dest = "cardinality", default = "16",
        help="Cardinality", type=int)
    parser.add_argument("-w", "--windows", dest = "window",
        default = "16", help="Sliding Window Size",
        type=int)
    parser.add_argument("-t", "--threshold",
        dest = "threshold", default = "1000",
        help="Threshold for split", type=int)
    parser.add_argument("-p", "--promotion",
        action='store_true',
        help="Define Promotion Strategy")
    parser.add_argument("TSfile")

    args = parser.parse_args()
```

This first part is about the `import` statements and reading the required parameters using `argparse`.

The second part of `createiSAX.py` is the following:

```
variables.segments = args.segments
variables.maximumCardinality = args.cardinality
variables.slidingWindowSize = args.window
variables.threshold = args.threshold
variables.defaultPromotion = args.promotion
file = args.TSfile

maxCardinality = variables.maximumCardinality
segments = variables.segments
windowSize = variables.slidingWindowSize

if tools.power_of_two(maxCardinality) == -1:
    print("Not a power of 2:", maxCardinality)
    sys.exit()

if variables.segments > variables.slidingWindowSize:
    print("Segments:", variables.segments,
        "Sliding window:", variables.slidingWindowSize)
    print("Sliding window size should be bigger than #
        of segments.")
    sys.exit()

print("Max Cardinality:", maxCardinality, "Segments:",
    variables.segments,
    "Sliding Window:", variables.slidingWindowSize,
    "Threshold:", variables.threshold,
    "Default Promotion:", variables.defaultPromotion)
```

In this part of `createiSAX.py`, we assign the parameters to the relevant local and global variables and make some tests to make sure that the parameters make sense. The reason for using local variables is to have smaller variable names to work with. The `print()` statement outputs the parameters on the screen.

The third part of `createiSAX.py` contains the following code:

```
ts = pd.read_csv(file, names=['values'],
    compression='gzip')
ts_numpy = ts.to_numpy()
length = len(ts_numpy)

#
```

```
# Initialize iSAX index
#
ISAX = isax.iSAX()
```

In this part, we read the compressed time series file and create the NumPy variable that holds the entire time series. After that, we initialize a variable to hold the iSAX index. As the name of the class is iSAX, the relevant variable is initialized as an instance of the isax.iSAX() class.

The last part of createiSAX.py contains the following code:

```
# Split sequence into subsequences
for i in range(length - windowSize + 1):
    # Get the subsequence
    ts = ts_numpy[i:i+windowSize]
    # Create new TS node based on ts
    ts_node = isax.TS(ts, segments)
    ISAX.insert(ts_node)

if __name__ == '__main__':
    main()
```

This last part splits the time series according to the sliding window size, creates the TS() objects, and inserts them into the iSAX index using the insert() method of the iSAX class using the ISAX variable – remember that it is iSAX.insert() that calls Node.insert().

Running createiSAX.py produces the following output:

```
$ ./createiSAX.py ts1.gz
Max Cardinality: 16 Segments: 16 Sliding Window: 16 Threshold: 1000
Default Promotion: False
$ ./createiSAX.py
usage: createiSAX.py [-h] [-s SEGMENTS] [-c CARDINALITY] [-w WINDOW]
[-t THRESHOLD] [-p] TSfile
createiSAX.py: error: the following arguments are required: TSfile
```

The good thing is that createiSAX.py has default values for all iSAX parameters. However, providing the path for the file that holds the time series is required.

In the next subsection, we are going to develop a command-line utility that counts the total number of subsequences in an iSAX index.

Counting the subsequences of an iSAX index

This is a really handy utility that not only shows *how to traverse an entire iSAX index* but also allows you to count all the subsequences of an iSAX index and make sure that you have not missed any subsequences in the process, which can be used for testing purposes.

The code of countSub.py that does the counting is the following – the rest of the implementation is the same as in createiSAX.py:

```
# Visit all entries in Dictionary
# Count TS in Terminal Nodes
sum = 0
for k in ISAX.ht:
    t = ISAX.ht[k]
    if t.terminalNode:
        sum += t.nTimeSeries()


print(length - windowSize + 1, sum)
```

The code visits the ISAX.ht field of an iSAX class because this is where *all the nodes* of the iSAX index are kept. If we are working with a terminal node, then we call the nTimeSeries() method to find the number of subsequences that are stored in that terminal node. We do that for all the terminal nodes, and we are done. The last statement prints both the theoretical number of subsequences as well as the actual number of subsequences found in the iSAX index. As long as these two values are the same, we are good.

Running countSub.py on a small time series, which you can find in the ch04 directory, generates the following kind of output:

```
$ ./countSub.py ts1.gz
Max Cardinality: 16 Segments: 16 Sliding Window: 16 Threshold: 1000
Default Promotion: False
35 35
```

The subsection that follows shows the time it takes to construct an iSAX index.

How long does it take to create an iSAX index?

In this subsection, we are going to compute the time it takes a computer to create an iSAX index. The main reason for any delays in the construction phase of iSAX is the splitting of nodes and the rearrangement of the subsequences. The more extensive splitting we have, the more time it takes to generate the index.

The code of howMuchTime.py that computes the time it takes to create the iSAX index is the following – the rest of the implementation is the same as in createiSAX.py:

```
start_time = time.time()
print("--- %.5f seconds ---" % (time.time() -
    start_time))
```

The first statement is located right before we begin reading the time series file using pd.read_csv() and the second statement is right after we finish splitting and inserting the time series into the iSAX index.

The output of howMuchTime.py when processing ts1.gz is similar to the following:

```
$ ./howMuchTime.py -w 2 -s 2 ts1.gz
Max Cardinality: 16 Segments: 2 Sliding Window: 2 Threshold: 1000
Default Promotion: False
--- 0.00833 seconds ---
```

As ts1.gz is a small time series with 50 elements, the output is not that interesting. Therefore, let us try using howMuchTime.py on bigger time series.

The next output shows the time it took howMuchTime.py to create an iSAX index for a time series with 500,000 elements on a macOS machine with 32 GB of RAM and an Apple M1 Max CPU – you can create a time series with the same length on your own and try the same command or use the provided file, which is called 500k.gz:

```
$ ./howMuchTime.py 500k.gz
Max Cardinality: 16 Segments: 16 Sliding Window: 16 Threshold: 1000
Default Promotion: False
--- 114.80277 seconds ---
```

The 500k.gz file was created using the following commands:

```
$ ./ch01/synthetic_data.py 500000 -1 1 > 500k
$ gzip 500k
```

The following output shows the time it took howMuchTime.py to create an iSAX index for a time series with 2,000,000 elements on a macOS machine with 32 GB of RAM and an Apple M1 Max CPU – you can create a time series with the same or a bigger length on your own and try the same command or use the provided file, which is called 2M.gz:

```
$ ./howMuchTime.py 2M.gz
Max Cardinality: 16 Segments: 16 Sliding Window: 16 Threshold: 1000
Default Promotion: False
--- 450.37358 seconds ---
```

The 2M.gz file was created using the following commands:

```
$ ./ch01/synthetic_data.py 2000000 -10 10 > 2M
$ gzip 2M
```

One interesting conclusion that we can make is that for a time series that is four times bigger, it took our program about four times longer to build. However, this is not always the case.

Additionally, the time it took to create an iSAX index does not tell the whole story, especially when testing on a busy machine or a slow machine with a little amount of RAM. What is more important is the number of node splits as well as the number of accesses to subsequences that were made. The minimum number of accesses to subsequences is equal to the subsequences of the time series. However, when splits take place, we must revisit the involved subsequences in order to distribute them according to the newly created SAX representations and terminal nodes. The splits and the revisiting of subsequences increase the construction time of the iSAX.

Therefore, we are going to create a modified version of howMuchTime.py that also prints the number of node splits, as well as the total number of subsequence accesses. The name of the new utility is accessSplit.py. The statements that do the counting of the splits and the subsequence accessing are already present in isax/isax.py and we just need to access two global variables, which are variables.nSplits and variables.nSubsequences, to get the results.

Running accessSplit.py with the default parameters on 500k.gz produces the following kind of output:

```
$ ./accessSplit.py 500k.gz
Max Cardinality: 16 Segments: 16 Sliding Window: 16 Threshold: 1000
Default Promotion: False
Number of splits: 0
Number of subsequence accesses: 499985
```

What does this output tell us? It tells us that there was no split! In practice, this means that the root node of that particular iSAX index has terminal nodes only as children. Is this good or bad? In general, this means that *the index works like a hash table* where the hash function is the function for calculating the SAX representation. Most of the time, this is not the desirable form that we want an index to have because we could have used a hash table in the first place!

If we run accessSplit.py on the same time series using different parameters, we are going to get a totally different output regarding the construction of the iSAX index:

```
$ ./accessSplit.py -w 1024 -s 8 -c 32 500k.gz
Max Cardinality: 32 Segments: 8 Sliding Window: 1024 Threshold: 1000
Default Promotion: False
Number of splits: 4733
Number of subsequence accesses: 16370018
```

What does this output tell us? It tells us that even on a relatively small time series, the iSAX parameters play a huge role in the creation time of the iSAX index. However, the number of subsequence accesses is around 33 times the total number of subsequences of the time series, which is pretty big and therefore not very efficient.

Let us now try accessSplit.py on a bigger time series, which is 2M.gz, and see what happens:

```
$ ./accessSplit.py 2M.gz
Max Cardinality: 16 Segments: 16 Sliding Window: 16 Threshold: 1000
Default Promotion: False
```

```
Number of splits: 0
Number of subsequence accesses: 1999985
```

As before, we are using the iSAX index as a hash table, which is not the desired behavior. Let us try using different parameters instead:

```
$ ./accessSplit.py -s 8 -c 32 2M.gz
Max Cardinality: 32 Segments: 8 Sliding Window: 16 Threshold: 1000
Default Promotion: False
Number of splits: 3039
Number of subsequence accesses: 13694075
```

This time, the number of accesses to subsequences is around seven times the length of the time series, which is more realistic than when we were working with the 500k.gz file.

We are going to work with accessSplit.py again in *Chapter 5*. But for now, we are going to learn more about the overflow of iSAX indexes.

Dealing with iSAX overflows

In this subsection, we are going to experiment with an overflow situation. Keep in mind that a dedicated global parameter exists in variables.py that holds the number of subsequences that were ignored due to overflow. Among other things, this helps you fix the iSAX parameters faster as you know how bad the overflow is. Usually, the easiest way to fix an overflow is by increasing the threshold value, but this might have serious implications when searching an iSAX index or comparing an iSAX index with another one.

The Python code of overflow.py has only one change compared to createiSAX.py, which is the following statement because the SAX representation that created the overflow is printed by default:

```
print("Number of overflows:", variables.overflow)
```

This mainly happens because the functionality is built into the isax package, which automatically prints a message when the first overflow takes place, and we just have to access the variables.overflow variable to find out the total number of overflows.

The output of overflow.py when working with the 500k.gz time series includes the following information:

```
$ ./overflow.py -w 1024 -s 8 500k.gz
Max Cardinality: 16 Segments: 8 Sliding Window: 1024 Threshold: 1000
Default Promotion: False
OVERFLOW: 1000_0111_0111_1000_1000_0111_0111_1000
Number of overflows: 303084
```

The previous output tells us that the first SAX representation that caused an overflow was `1000_0` `111_0111_1000_1000_0111_0111_1000` and that we had `303084` overflows in total – we might have more SAX representations that cause an overflow, but we have decided to print just the first one. This means that `303084` subsequences were not inserted into the iSAX index, which is a very large number compared to the length of the time series.

Let us now try the same command using the other promotion strategy and see what happens:

```
$ ./overflow.py -w 1024 -s 8 500k.gz -p
Max Cardinality: 16 Segments: 8 Sliding Window: 1024 Threshold: 1000
Default Promotion: True
Non recoverable Promotion overflow!
OVERFLOW: 1000_0111_0111_1000_1000_0111_0111_1000
Number of overflows: 303084
```

It turns out that we got the same kind of overflow and the exact same number of total overflows. This makes perfect sense as the overflow situation does not have to do with the promotion strategy but with the SAX representations. *A different promotion strategy might change the shape of an iSAX index a little, but it has nothing to do with the overflow situation.*

As `303084` is a big number, we might need to drastically increase the capacity of the iSAX index but without creating an unnecessarily big iSAX index. So, with that in mind, we can try to resolve the overflow situation by changing the parameters of the iSAX index. Let us try to do so then by increasing the threshold value:

```
$ ./overflow.py -w 1024 -s 8 -c 16 -t 1500 500k.gz
Max Cardinality: 16 Segments: 8 Sliding Window: 1024 Threshold: 1500
Default Promotion: False
OVERFLOW: 0111_1000_1000_1000_1000_0111_0111_0111
Number of overflows: 176454
```

So, it looks like we cut the number of overflows in half, which is a good thing for a start. However, although we are using the same cardinality as before, it is a different SAX representation (`0111_100` `0_1000_1000_1000_0111_0111_0111`) that caused the first overflow this time, which means that the increased threshold value solved the overflow condition that took place earlier with the `1000` `_0111_0111_1000_1000_0111_0111_1000` SAX representation.

Let us give it another try by increasing the `cardinality` value and at the same time decreasing the threshold value:

```
$ ./overflow.py -w 1024 -s 8 -c 32 -t 500 500k.gz
Max Cardinality: 32 Segments: 8 Sliding Window: 1024 Threshold: 500
Default Promotion: False
Number of overflows: 0
```

So, we finally found a triplet of parameters that works for the 1024 sliding window size and the 500k.gz dataset.

Is there a recipe for finding out which parameters work and which do not work? No, as it mainly depends on the values of the dataset and the sliding window size. The more you work and experiment with iSAX indexes, the more you are going to understand which parameters work best for a given dataset and sliding window size.

So, in this last section, we learned about iSAX overflows and presented a technique for solving such situations.

Summary

In this chapter, we saw and went through the implementation details of the isax Python package, which allows us to create iSAX indexes. Please make sure that you understand the code and most importantly know how to use the code.

Additionally, we implemented many command-line utilities that allow us to create iSAX indexes and understand what happens behind the scenes regarding the splits and subsequence accesses as well as the overflow conditions. Having a better understanding of the structure of an iSAX index allows us to select better indexes and avoid using poor ones.

The next chapter is going to put iSAX indexes into practice by showing how to search and join iSAX indexes.

Useful links

- The argparse package: https://docs.python.org/3/library/argparse.html
- The NumPy Python package: https://numpy.org/

Exercises

Try to work through the following exercises:

- Create a synthetic dataset with 100,000 elements with values from *-10 to 10* and construct an iSAX index with 4 segments, a cardinality of 64, and a threshold value of 1000. How much time did it take your machine to create the iSAX index? Are there any overflows?

- Create a synthetic dataset with 100,000 elements with values from *-1 to 1* and construct an iSAX index with 4 segments, a cardinality of 64, and a threshold value of 1000. How much time did it take your machine to create that iSAX index?

- Create a synthetic dataset with 500,000 elements with values from *0 to 10* and construct an iSAX index with 4 segments, a cardinality of 64, and a threshold value of `1000`. How much time did it take your machine to create the iSAX index?

- Create a synthetic dataset with 500,000 elements with values from *0 to 10* and construct an iSAX index with 4 segments, a cardinality of 64, and a threshold value of `1000`. How many splits and accesses to subsequences took place? What happens if you increase the threshold value to `1500`?

- Create a synthetic dataset with 150,000 elements with values from *-1 to 1* and construct an iSAX index with 4 segments, a cardinality of 64, and a threshold value of `1000`. Are there any overflows? How many splits were performed for the construction of the iSAX index?

- Experiment with `accessSplit.py` on `2M.gz` using various iSAX parameters. Which parameters seem to work best? Do not forget that high threshold values have a great impact on searching; so, in general, do not use huge threshold values to lower the number of splits.

- Experiment with `accessSplit.py` on `500k.gz` using various iSAX parameters. Which parameters seem to work best?

5

Joining and Comparing iSAX Indexes

In the previous chapter, we developed a Python package called `isax` that creates iSAX indexes for indexing the subsequences of a time series given a sliding window.

In this chapter, we are going to experiment with how the sliding window size affects the number of splits and the number of accesses to subsequences while creating an iSAX index.

Then, we are going to use the iSAX indexes created by the `isax` package and try to join and compare them. By *comparing*, we aim to understand the efficiency of an iSAX index, and by *joining*, we mean being able to find similar nodes in two iSAX indexes based on SAX representations.

The last part of this chapter is going to briefly discuss Python testing before developing simple tests for the `isax` package. *Testing is a serious part of the development process and should not be overlooked. The time spent writing tests is time well spent!*

In this chapter, we are going to cover the following main topics:

- How the sliding window size affects the iSAX construction speed
- Checking the search speed of iSAX indexes
- Joining iSAX indexes
- Implementing the joining of iSAX indexes
- Explaining the Python code
- Using the Python code
- Writing Python tests

Technical requirements

The GitHub repository for this book can be found at `https://github.com/PacktPublishing/Time-Series-Indexing`. The code for each chapter is in its own directory. Therefore, the code for this chapter can be found in the `ch05` folder. You can download the entire repository on your computer using `git(1)` or you can access the desired files via the GitHub user interface.

How the sliding window size affects the iSAX construction speed

In this section, we are going to continue working with the `accessSplit.py` utility we developed in the previous chapter to find out whether the sliding window size affects the construction speed of an iSAX index, provided that the remaining iSAX parameters stay the same.

Put simply, we will use different methods to find out more about the quality of iSAX indexes and whether the sliding window size affects the construction speed. We are going to perform our experiments using the following sliding window sizes: 16, 256, 1024, 4096, and 16384. We are going to experiment using the $500k.gz$ time series from *Chapter 4*, 8 segments, a maximum cardinality value of 32, and a threshold value of 500.

For the window size of 16, the results are the following:

```
$ ./accessSplit.py -s 8 -c 32 -t 500 -w 16 500k.gz
Max Cardinality: 32 Segments: 8 Sliding Window: 16 Threshold: 500
Default Promotion: False
Number of splits: 1376
Number of subsequence accesses: 2776741
```

For the sliding window size of 256, the results are the following:

```
$ ./accessSplit.py -s 8 -c 32 -t 500 -w 256 500k.gz
Max Cardinality: 32 Segments: 8 Sliding Window: 256 Threshold: 500
Default Promotion: False
Number of splits: 4234
Number of subsequence accesses: 10691624
```

Compared to the sliding window size of 16, the iSAX index created using a sliding window size of 256 had more than three times the number of splits and four times the number of subsequence accesses.

Next, for the window size of 1024, the results are the following:

```
$ ./accessSplit.py -s 8 -c 32 -t 500 -w 1024 500k.gz
Max Cardinality: 32 Segments: 8 Sliding Window: 1024 Threshold: 500
Default Promotion: False
Number of splits: 5983
Number of subsequence accesses: 15403024
```

As before, we have more splits than the 16 and 256 sliding window sizes and more subsequence accesses. Put simply, it took more CPU time for this iSAX index to be constructed.

Next, for the window size of 4096, the results are the following:

```
$ ./accessSplit.py -s 8 -c 32 -t 500 -w 4096 500k.gz
Max Cardinality: 32 Segments: 8 Sliding Window: 4096 Threshold: 500
Default Promotion: False
OVERFLOW: 10000_10000_01111_01111_01111_10000_10000_01111
Number of splits: 6480
Number of subsequence accesses: 18537820
```

In this case, it is not only slower to construct the iSAX index but we also have an **overflow situation**. Therefore, the subsequences of the 500.gz time series are not going to fit into an iSAX index with these parameters and we are going to need to use different iSAX parameters for the iSAX index to work.

Do overflows have an impact on the construction of iSAX indexes?

When we have one or multiple overflows on an iSAX index, it means that the full cardinality has been used on all SAX words—recall that the number of SAX words is defined by the number of segments. Therefore, we have multiple splits on terminal nodes that are fully based on the current threshold value, which means that we have many more subsequence accesses than usual. Therefore, overflows have a great impact on the construction time of iSAX indexes. Additionally, as if this were not bad enough, we have to find new iSAX parameters that prevent the overflow from happening while keeping the iSAX operation efficient. Keep in mind that the number of splits is also a naïve indication of how close we are to an overflow.

Lastly, for the biggest window size (16384), the results are the following:

```
$ ./accessSplit.py -s 8 -c 32 -t 500 -w 16384 500k.gz
Max Cardinality: 32 Segments: 8 Sliding Window: 16384 Threshold: 500
Default Promotion: False
OVERFLOW: 01111_10000_10000_01111_10000_01111_10000_01111
Number of splits: 6996
Number of subsequence accesses: 19201125
```

Once again, we have an overflow situation with the sliding window size of 16384, this time on a different SAX representation. We are going to leave both overflows as they are and create some plots of the results. The resolution of the overflows is left as an exercise for you.

Figure 5.1 shows the number of splits per sliding window size where we can see that the bigger the sliding window size, the larger the number of splits that take place for that particular time series.

Splits versus Sliding Window

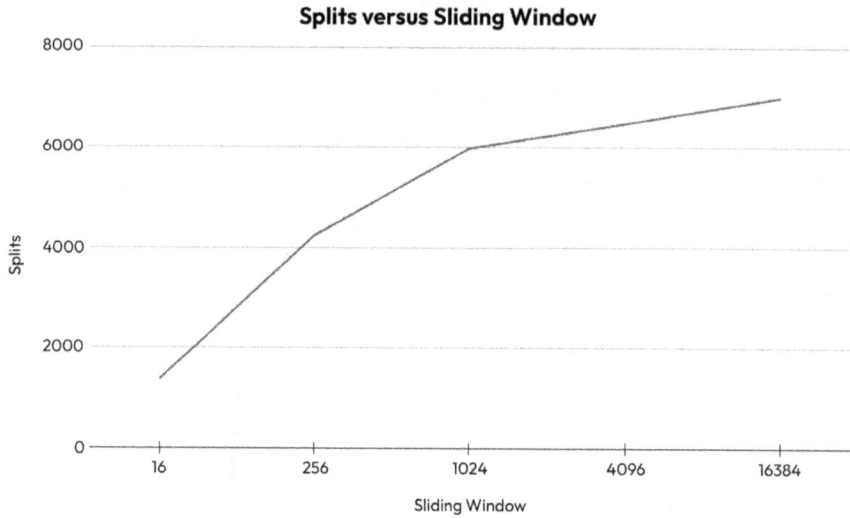

Figure 5.1– Splits per sliding window size plot

Figure 5.2 shows the number of subsequence accesses per sliding window size. In this case, instead of plotting the absolute number of subsequence accesses, we divide the number of total subsequence accesses by the total number of subsequences to display a fraction. This is a fair calculation as bigger time series have a larger number of subsequences.

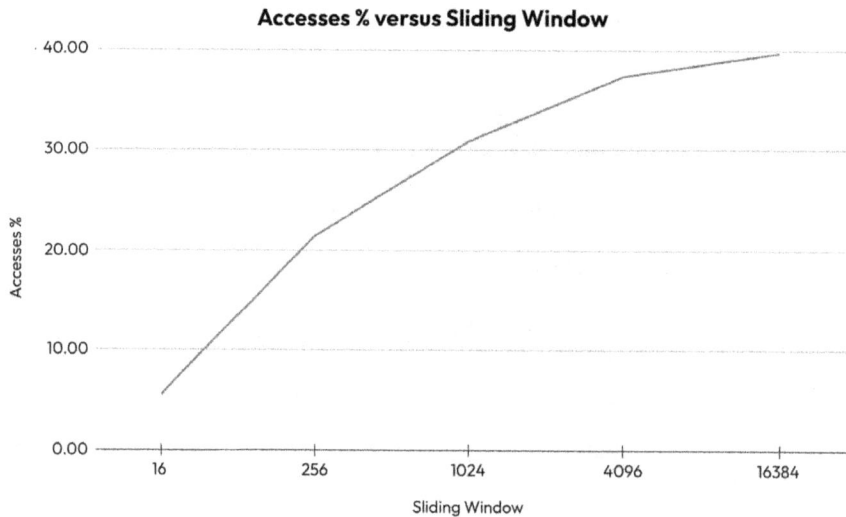

Accesses % versus Sliding Window

Figure 5.2 – Subsequence accesses percentage per sliding window size plot

In *Figure 5.2*, we can see that the bigger the sliding window size, the bigger the number of subsequence accesses. For the smallest sliding window (16), we have about eight times fewer accesses to the subsequence of the time series compared to the largest sliding window (16384).

The construction speed of an iSAX index is one important factor. However, it is not the only criterion for the quality of iSAX indexes. The next section investigates the search speed of iSAX indexes.

Checking the search speed of iSAX indexes

This section presents a utility that takes two time series, named TS1 and TS2, which ideally have similar lengths, creates two iSAX indexes, named D1 and D2, and performs the following searches:

- Searches D1 for all the subsequences of TS2. In this case, we are not sure whether a subsequence from TS2 is in D1 or not. In most cases, *we are not going to be able to find the subsequences of TS2 in TS1*. This is the main reason that a *join* based on the SAX representations of the iSAX nodes might be more appropriate when looking for similarities between subsequences.

- Searches D2 for all the subsequences of TS1. In this case, we are not sure whether a subsequence from TS1 is in D2 or not. As before, in most cases, we are not going to be able to find the subsequences of TS1 in TS2 and therefore, in the iSAX index created from TS2 (D2).

- Searches D1 for all the subsequences of TS1, which means that all subsequences of TS1 are in D1. With that test, we just want to discover more about the speed of an iSAX index when performing search operations. This search operation mainly depends on the threshold size because a bigger threshold value means more subsequences to look for once we come to the appropriate terminal node.

- Searches D2 for all the subsequences of TS2, which means that all subsequences of TS2 are in D2 and are going to be found.

All these searches are implemented in a Python script named speed.py.

The core functionality of speed.py is implemented in functions. The first function contains the following code:

```python
def createISAX(file, w, s):
    # Read Sequence as Pandas
    ts = pd.read_csv(file, names=['values'],
        compression='gzip').astype(np.float64)

    # Convert to NParray
    ts_numpy = ts.to_numpy()
    length = len(ts_numpy)

    ISAX = isax.iSAX()
    ISAX.length = length
```

```
# Split sequence into subsequences
for i in range(length - w + 1):
    # Get the subsequence
    ts = ts_numpy[i:i+w]
    # Create new TS node based on ts
    ts_node = isax.TS(ts, s)
    ISAX.insert(ts_node)

return ISAX, ts_numpy
```

The `createISAX()` function creates an iSAX index and returns a link to an `isax.ISAX()` class as well as a NumPy array with all the elements of the time series.

The second function is implemented as follows:

```
def query(ISAX, q):
    global totalQueries
    totalQueries = totalQueries + 1
    Accesses = 0

    # Create TS Node
    qTS = isax.TS(q, variables.segments)

    segs = [1] * variables.segments
    #If the relevant child of root is not there, we have a miss
    lower_cardinality = tools.lowerCardinality(segs, qTS)

    lower_cardinality_str = ""
    for i in lower_cardinality:
        lower_cardinality_str = lower_cardinality_str + "_"
            + i
```

In the first part of `query()`, we construct the SAX representation of a potential child of the root node of the iSAX index using `tools.lowerCardinality()` and `segs`. From that, we construct the `lower_cardinality_str` string:

```
# Remove _ at the beginning
Lower_cardinality_str = lower_cardinality_str[1:len(
    lower_cardinality_str)]
if ISAX.ht.get(lower_cardinality_str) == None:
    return False, 0

# Otherwise, we have a hit
n = ISAX.ht.get(lower_cardinality_str)
while n.terminalNode == False:
```

```
        left = n.left
        right = n.right

        leftSegs = left.word.split('_')
        # Promote
        tempCard = tools.promote(qTS, leftSegs)

        if tempCard == left.word:
            n = left
        elif tempCard == right.word:
            n = right

    # Iterate over the subsequences of the terminal node
    for i in range(0, variables.threshold):
        Accesses = Accesses + 1
        child = n.children[i]
        if type(child) == isax.TS:
            # print("Shapes:", child.ts.shape, qTS.ts.shape)
            if np.allclose(child.ts, qTS.ts):
                return True, Accesses
        else:
            return False, Accesses

    return False, Accesses
```

In the second part of `query()`, we check whether the `lower_cardinality_str` key can be found in the iSAX index.

If it can be found, then we follow that subtree, which begins with a child of the root node of the iSAX index, until we find the appropriate terminal node. If it cannot be found, then we have a miss, and the process terminates.

The `query()` function returns `True` if the subsequence is found and `False` otherwise. Its second return value is the number of subsequence accesses that took place while trying to find that query subsequence.

The rest of the code of `speed.py` is placed in the `main()` function and is going to be presented in three parts – the first part being the following:

```
# Build iSAX for TS1
i1, ts1 = createISAX(f1, windowSize, segments)
totalSplits = totalSplits + variables.nSplits
totalAccesses = totalAccesses + variables.nSubsequences

# Build iSAX for TS2
variables.nSubsequences = 0
```

```
variables.nSplits = 0
i2, ts2 = createISAX(f2, windowSize, segments)
totalSplits = totalSplits + variables.nSplits
totalAccesses = totalAccesses + variables.nSubsequences
```

In this first part, we construct the two iSAX indexes and store the number of splits and accesses to subsequences.

The second part of speed.py contains the following code:

```
# Query iSAX for TS1
for idx in range(0, len(ts1)-windowSize+1):
    currentQuery = ts1[idx:idx+windowSize]
    found, ac = query(i1, currentQuery)
    if found == False:
        print("This cannot be happening!")
        return
    totalAccesses = totalAccesses + ac

# Query iSAX for TS1
for idx in range(0, len(ts2)-windowSize+1):
    currentQuery = ts2[idx:idx+windowSize]
    found, ac = query(i1, currentQuery)
    totalAccesses = totalAccesses + ac
```

In this part of the program, we query the first iSAX index. In the first for block, we search iSAX for all the subsequences of the first time series. As this iSAX indexes the first time series, all subsequences are going to be found in the iSAX index. While doing that, we store the number of accesses to subsequences, which is returned by the query() function. In the second for block, we do the same but this time, for the second time series. Therefore, there is a small possibility of finding the subsequences of the second time series (TS2) in the iSAX index of the first time series (TS1).

The last part of speed.py is as follows:

```
# Query iSAX for TS2
for idx in range(0, len(ts2)-windowSize+1):
    currentQuery = ts2[idx:idx+windowSize]
    found, ac = query(i2, currentQuery)
    if found == False:
        print("This cannot be happening!")
        return
    totalAccesses = totalAccesses + ac

# Query iSAX for TS2
for idx in range(0, len(ts1)-windowSize+1):
    currentQuery = ts1[idx:idx+windowSize]
```

```
        found, ac = query(i2, currentQuery)
        totalAccesses = totalAccesses + ac
```

This last part of main() is similar to the previous code. The only difference is that this time, we query the second iSAX index instead of the first one. Once again, we store the number of accesses to subsequences.

Before running speed.py, we need to create another time series, which is going to be stored in 506k.gz. In this case, the second time series was created as follows:

```
$ ../ch01/synthetic_data.py 506218 -10 10 > 506k
$ gzip 506k
```

Although the two time series do not need to have the same length, we have decided to make them pretty close in length.

Using speed.py generates the following kind of output:

```
$ ./speed.py -s 8 500k.gz 506k.gz
Max Cardinality: 16 Segments: 8 Sliding Window: 16 Threshold: 1000
Default Promotion: False
Total subsequence accesses: 1060326778
Total splits: 1106
Total queries: 2012376
```

Keep in mind that the previous command took *more than three hours* on a MacBook Pro machine! The speed will depend on your CPU.

If we used different SAX parameters, the output would look as follows:

```
$ ./speed.py -s 4 -c 64 500k.gz 506k.gz
Max Cardinality: 64 Segments: 4 Sliding Window: 16 Threshold: 1000
Default Promotion: False
Total subsequence accesses: 1083675402
Total splits: 2034
Total queries: 2012376
```

Although the first run of speed.py required 1,106 splits and the second one 2,034 splits, both results were close as far as the total number of subsequence accesses is concerned.

As expected, the number of total queries is the same in both cases because we are dealing with the same time series and, therefore, the same number of subsequences.

Now that we know how to perform lookups and searches on iSAX indexes, it is time to learn about another important operation, which is the joining of iSAX indexes.

Joining iSAX indexes

At this point, we have iSAX indexes that we want to use to perform basic time series data mining tasks. One of them is **finding similar subsequences** between two or more time series. In our case, we are working with two time series, but the method can be extended to more time series with small changes.

> **How to join iSAX indexes**
>
> Given two or more iSAX indexes, it is up to us to decide how and why we are going to join them. We can even join them using SAX representations with a cardinality value of 2. However, using the SAX representations of the nodes as our keys for the join is the most logical choice. In our case, we are going to use the iSAX indexes and the SAX representations of the nodes to look for similar subsequences. This is because we have the intuition that subsequences in nodes with the same SAX representation are *close* to each other. The term *close* is defined relative to a **distance metric**. For the purposes of this chapter, we are going to use the Euclidean distance to compare subsequences of the same size.

Now, let us rephrase that in a more precise way. A **join** of two iSAX indexes that is based on the SAX representation is a way of finding the most similar node (based on the SAX representation) for each of the nodes of the first iSAX index when searching the nodes of the second iSAX index, which was constructed using the same parameters. This way, we save time because we only have to compare subsequences of similar terminal nodes. Is similarity based on SAX representation perfect? No, it is not. But we are using a time series index to make things faster.

The idea behind this join came after reading the *Scalable Hybrid Similarity Join over Geolocated Time Series* paper, which was written by Georgios Chatzigeorgakidis, Kostas Patroumpas, Dimitrios Skoutas, Spiros Athanasiou, and Spiros Skiadopoulos.

The next section is going to show how to implement the joining of iSAX indexes based on the SAX representations of their nodes.

Implementing the joining of iSAX indexes

For the implementation of the iSAX index join, we are going to assume that we have two iSAX indexes ready to be used saved in two separate Python variables, and continue from there. We are going to need a Python function that accepts two iSAX indexes and returns a list of Euclidean distances, which are the nearest neighbors of all subsequences in both time series. Keep in mind that if a node from one of the iSAX indexes does not match the other iSAX index, then that node, and as a consequence its subsequences, is not going to get processed. Therefore, the list of Euclidean distances might be a little shorter than expected. That is the main reason why we must not use unnecessarily big iSAX parameters. Put simply, do not use 16 segments when 4 segments can do the job.

Additionally, keep in mind that the real nearest neighbor of a subsequence might not be located in the terminal node with the same SAX representation – this is the price we pay for the extra speed and avoiding the quadratic processing cost (comparing every subsequence of the first time series with all subsequences of the second time series, and vice versa).

So, we need to implement the previous functionality for the isax package based on the current implementation and representation of the iSAX index.

As a result, we are going to put that functionality inside the isax package, using a separate file named iSAXjoin.py.

Apart from that file, we added a function for calculating the Euclidean distance between two subsequences in isax/tools.py:

```
def euclidean(a, b):
    return np.linalg.norm(a-b)
```

If you recall from *Chapter 1*, in the ch01/ed.py script, euclidean() uses the magic of NumPy to calculate Euclidean distances between two subsequences. Do not forget that we always compare **normalized subsequences** in this book.

Lastly, we added the following variable to isax/variables.py:

```
# List of Euclidean distances
ED = []
```

The ED global variable is a Python list used to hold the result of the join between two iSAX indexes, which is a list of Euclidean distances.

Let us now present and explain the code of isax/iSAXjoin.py.

Explaining the Python code

The code in iSAXjoin.py is going to be presented in five parts.

The first part is the following:

```
from isax import variables
from isax import tools

def Join(iSAX1, iSAX2):
    # Begin with the children of the root node.
    # That it, the nodes with SAX words
    # with a Cardinality of 1.
    for t1 in iSAX1.children:
        k1 = iSAX1.children[t1]
        if k1 == None:
            continue
        for t2 in iSAX2.children:
```

```
            k2 = iSAX2.children[t2]
            if k2 == None:
                continue
            # J_AB
            _Join(k1, k2)
            # J_BA
            _Join(k2, k1)
    return
```

The Join() function is the entry point to the join of the two iSAX indexes. However, that function has a single purpose, which is creating all the combinations between the children of the two iSAX root nodes in order to pass control to _Join(). As the order of the arguments in _Join() is significant, _Join() is called two times. The first time, a root child node from the first iSAX index is the first parameter, and the second time, a root child node from the second iSAX index is the first parameter.

The second part of iSAXjoin.py is as follows:

```
def _Join(t1, t2):
    if t1.word != t2.word:
        return
    # Inner + Inner
    if t1.terminalNode==False and t2.terminalNode==False:
        _Join(t1.left, t2.left)
        _Join(t1.right, t2.left)
        _Join(t1.left, t2.right)
        _Join(t1.right, t2.right)
```

When we are dealing with inner nodes from both iSAX indexes, we just combine all their children – remember that each inner node has two children – and recursion takes care of the rest.

The third part contains the following code:

```
    # Terminal + Inner
    elif t1.terminalNode==True and t2.terminalNode==False:
        _Join(t1, t2.left)
        _Join(t1, t2.right)
```

If we are dealing with an inner node and a terminal node, we expand the inner node, and recursion takes care of the rest.

The fourth part of iSAXjoin.py is as follows:

```
    # Inner + Terminal
    elif t1.terminalNode == False and t2.terminalNode == True:
        _Join(t1.left, t2)
        _Join(t1.right, t2)
```

As before, when dealing with an inner node and a terminal node, we expand the inner node, and recursion takes care of the rest.

The last part comes with the following Python code:

```
    # Terminal + Terminal
    # As both are terminal nodes, calculate
    # Euclidean Distances between Time Series pairs
    elif t1.terminalNode==True and t2.terminalNode==True:
        for i in range(t1.nTimeSeries()):
            minDistance = None
            for j in range(t2.nTimeSeries()):
                distance =round(tools.euclidean
                (t1.children[i].ts, t2.children[j].ts),
                variables.precision)
                # Keeping the smallest Euclidean Distance for each
  node
                # of the t1 Terminal node
                if minDistance == None:
                    minDistance = distance
                elif minDistance > distance:
                    minDistance = distance
            # Insert distance to PQ
            if minDistance != None:
                variables.ED.append(minDistance)
    else:
        print("This cannot happen!")
```

This last part is where the recursive calling to _Join() stops because we are dealing with two terminal nodes. This means that we are able to calculate the Euclidean distances of their subsequences. The fact that we are not normalizing the subsequences before calling tools.euclidean() means that we expect to have the subsequences in all terminal nodes in a normalized form. Notice that we store the results in the variables.ED list.

That is all regarding the implementation of the joining of two iSAX indexes. The next section covers how to use the (similarity) join code.

Using the Python code

In this section, we are going to use the similarity join code we have developed to start joining iSAX indexes. The source code of join.py is presented in three parts. The first part is the following:

```
#!/usr/bin/env python3

from isax import variables
from isax import isax
```

```python
from isax import tools
from isax.sax import normalize
from isax.iSAXjoin import Join

import sys
import pandas as pd
import time
import argparse

def buildISAX(file, windowSize):
    variables.overflow = 0

    # Read Sequence as Pandas
    ts = pd.read_csv(file, names=['values'],
        compression='gzip', header = None)

    ts_numpy = ts.to_numpy()
    length = len(ts_numpy)

    ISAX = isax.iSAX()
    ISAX.length = length

    for i in range(length - windowSize + 1):
        ts = ts_numpy[i:i+windowSize]
        # Create new TS node based on ts
        # Store the normalized version of the subsequence
        ts_node = isax.TS(normalize(ts),
            variables.segments)
        ISAX.insert(ts_node)

    if variables.overflow != 0:
        print("Number of overflows:", variables.overflow)

    return ISAX
```

Nothing new here – we just have to import the necessary external packages, including `isax.iSAXjoin`, and develop a function that creates an iSAX index given a time series file and a sliding window size. The function returns the root node of the iSAX index. However, please note that *subsequences are stored in their normalized form* inside all `TS()` objects.

The second part is the beginning of the `main()` function and comes with the following Python code:

```python
def main():
    parser = argparse.ArgumentParser()
    parser.add_argument("-s", "--segments",
        dest = "segments", default = "16",
```

```
        help="Number of Segments", type=int)
    parser.add_argument("-c", "--cardinality",
        dest = "cardinality", default = "256",
        help="Cardinality", type=int)
    parser.add_argument("-w", "--window", dest = "window",
        default = "16", help="Sliding Window Size",
        type=int)
    parser.add_argument("-t", "--threshold",
        dest = "threshold", default = "50",
        help="Threshold for split", type=int)
    parser.add_argument("-p", "--promotion",
        action='store_true',
        help="Define Promotion Strategy")
    parser.add_argument("TS1")
    parser.add_argument("TS2")


    args = parser.parse_args()


    variables.segments = args.segments
    variables.maximumCardinality = args.cardinality
    variables.slidingWindowSize = args.window
    variables.threshold = args.threshold
    variables.defaultPromotion = args.promotion


    windowSize = variables.slidingWindowSize
    maxCardinality = variables.maximumCardinality
    f1 = args.TS1
    f2 = args.TS2


    if tools.power_of_two(maxCardinality) == -1:
        print("Not a power of 2:", maxCardinality)
        sys.exit()


    if variables.segments > variables.slidingWindowSize:
        print("Segments:", variables.segments,
            "Sliding window:", variables.slidingWindowSize)
        print("Sliding window size should be bigger than # of
segments.")
        sys.exit()


    print("Max Cardinality:", maxCardinality, "Segments:",
        variables.segments,
        "Sliding Window:", variables.slidingWindowSize,
        "Threshold:", variables.threshold,
        "Default Promotion:", variables.defaultPromotion)
```

As we are joining two iSAX indexes, we need two separate command-line arguments (TS1 and TS2) to define the paths of the compressed plain text files with the time series data.

The last part of join.py is as follows:

```
# Build iSAX for TS1
start_time = time.time()
i1 = buildISAX(f1, windowSize)
print("i1: %.2f seconds" % (time.time() - start_time))

# Build iSAX for TS2
start_time = time.time()
i2 = buildISAX(f2, windowSize)
print("i2: %.2f seconds" % (time.time() - start_time))

# Join the two iSAX indexes
Join(i1, i2)
variables.ED.sort()
print("variables.ED length:", len(variables.ED))
maximumLength = i1.length+i2.length - 2*windowSize + 2
print("Maximum length:", maximumLength)

if __name__ == '__main__':
    main()
```

Here, we create the two iSAX indexes by calling buildISAX() two times and then join them using Join(), which returns no values. In order to look at the list of computed values, we need to access variables.ED. We print the length of the list of Euclidean distances as well as the theoretical maximum length of it, which is equal to time_series_length - sliding_window_size + 1, to have a better idea of the number of subsequences without a match. In the output, we also print the time it took to create each iSAX index as extra information about the process.

At this point, we are ready to use join.py. This means that we should provide it with the necessary parameters and input.

Using join.py produces the following kind of output:

```
$ ./join.py -s 8 -c 32 -t 1000 500k.gz 506k.gz
Max Cardinality: 32 Segments: 8 Sliding Window: 16 Threshold: 1000
Default Promotion: False
i1: 170.94 seconds
i2: 179.80 seconds
variables.ED length: 970603
Maximum length: 1006188
```

So, it took `170.94` seconds to create the first index and `179.80` seconds to create the second iSAX index. The list of Euclidean distances has `970603` elements, whereas the maximum number of elements is `1006188`, which means that we missed some terminal nodes because their SAX representation did not have a match in the other iSAX index. This is not unusual and we should expect it most of the time as time series and their iSAX indexes differ.

We have a long list of Euclidean distances, so what?

You might be asking, "What do we do with that list of Euclidean distances?" Put simply, what is the main purpose of creating such a list of distances? There are many uses, including the following:

- Finding out how close two time series are by finding the minimum Euclidean distance in the list.

- Finding out the list of Euclidean distances that are in a given numeric range. This is another way of comparing the similarity of two time series.

- Finding subsequences that are more different than others based on a distance measure. In data mining terminology, these subsequences are called **outliers**.

I think you get the idea of why we perform the join computation – we need to be able to better understand the connection between the two time series involved in the join. The reason for using the SAX representation is to prune nodes and subsequences and save CPU time.

As the joining operation can be slow, the next subsection presents a handy technique for saving the list of Euclidean distances on disk and loading the list from disk in order to use it without having to carry out the entire process from scratch.

Saving the output

The joining of iSAX indexes can take time. Is there a way to make that process less painful? Yes, we can save the contents of the similarity join, which is a list, into a file, which saves us from having to recreate that list from scratch each time we need it. Keep in mind that for this to work, the two iSAX indexes must be created with the same parameters for the exact same time series.

The `saveLoadList.py` script demonstrates the idea in the `main()` function – you can see the implementation of `buildISAX()` in `join.py`. The first part of `main()` is as follows. Some code is omitted for brevity:

```
def main():
    . . .
    # Reading command line parameters
    . . .

    # Build iSAX for TS1
    start_time = time.time()
```

```
i1 = buildISAX(f1, windowSize)
print("i1: %.2f seconds" % (time.time() - start_time))
# Build iSAX for TS2
start_time = time.time()
i2 = buildISAX(f2, windowSize)
print("i2: %.2f seconds" % (time.time() - start_time))


# Now, join the two iSAX indexes
Join(i1, i2)
variables.ED.sort()


print("variables.ED length:", len(variables.ED))


# Now save it to disk
#
# Define filename
filename = "List_" + basename(f1) + "_" + basename(f2) + "_"
+ str(maxCardinality) + "_" + str(variables.segments) + "_" +
str(windowSize) + ".txt"
print("Output file:", filename)
f = open(filename, "w")


# Write to disk
for item in variables.ED:
    f.write('%s\n' %item)


f.close()
```

In the previous code, we put the similarity join data into `variables.ED` by calling `Join()` from `isax.iSAXjoin` and printing its length. After that, we computed the filename of the output file that is saved in the `filename` variable, which is based on the parameters of the program. This is a handy way of creating **descriptive filenames** that reveal information about their components. After that, we wrote the contents of `variables.ED` into that file.

The second part of `main()` contains the following code:

```
# Now try to open it
f = open(filename, "r")

PQ = []
for item in f.readlines():
    PQ.append(float(item.rstrip()))

f.close()
print("PQ length:", len(PQ))
```

In the previous code, we tried to read the filename that we used for storing the contents of variables. ED and put the contents of the plain text file into the PQ variable. Lastly, we printed the length of PQ in order to compare it with the length of variables. ED and make sure that everything worked as expected.

Running the saveLoadList.py script generates the following output:

```
$ ./saveLoadList.py -s 8 -c 32 -t 1000 500k.gz 506k.gz
Max Cardinality: 32 Segments: 8 Sliding Window: 16 Threshold: 1000
Default Promotion: False
i1: 168.73 seconds
i2: 172.39 seconds
variables.ED length: 970603
Output file: List_500k.gz_506k.gz_32_8_16.txt
PQ length: 970603
```

From the previous output, we understand that the list contains 970603 elements. Additionally, the filename where we saved the contents of the list is List_500k.gz_506k.gz_32_8_16.txt. The only information missing from the filename is the threshold value.

The next subsection presents a utility that finds the nodes of an iSAX index that do not have a match in another iSAX index, and vice versa.

Finding iSAX nodes without a match

In this subsection, we specify the nodes of an iSAX index that do not have a match in another iSAX index, and vice versa. In reality, we are going to print the SAX representations of the terminal nodes of each iSAX index that do not have a match to a terminal node on the other iSAX.

The noMatch.py script implements the idea using the following code – we assume that we have already created two iSAX indexes for the two time series so that we do not have to repeat the code for creating iSAX indexes:

```
# Visit all entries in Dictionary
sum = 0
for k in i1.ht:
    t = i1.ht[k]
    if t.terminalNode:
        saxWord = t.word
        # Look for a match in the other iSAX
        if saxWord in i2.ht.keys():
            i2Node = i2.ht[saxWord]
            # Need that to be a terminal node
            if i2Node.terminalNode == False:
                sum = sum + 1
```

```
                print(saxWord, end=' ')

    print()
```

The previous code visits all the nodes of the first iSAX index looking for terminal nodes. Once a terminal node is found, we get its SAX representation and look in the other iSAX index for a terminal node with the same SAX representation. If such a node cannot be found, we print the SAX representation of the terminal node of the first iSAX index that does not have a match.

We should now use the same process for the second time series and the second iSAX index. The code presented here is similar to the previous one:

```
    # Look at the other iSAX
    for k in i2.ht:
        t = i2.ht[k]
        if t.terminalNode:
            saxWord = t.word
            # Look for a match in the other iSAX
            if saxWord in i1.ht.keys():
                i1Node = i1.ht[saxWord]
                # Sstill need that to be a terminal node
                if i1Node.terminalNode == False:
                    sum = sum + 1
                    print(saxWord, end=' ')

    print()
    print("Number of iSAX nodes without a match:", sum)
```

So, right after examining the second iSAX index, we print the total number of terminal nodes without a match.

Running noMatch.py generates the following kind of output:

```
$ ./noMatch.py -s 8 -c 32 -t 1500 -w 128 500k.gz 506k.gz
Max Cardinality: 32 Segments: 8 Sliding Window: 128 Threshold: 1500
Default Promotion: False
011_10_10_10_01_10_10_01 011_01_01_10_10_10_10_10
011_01_10_01_10_10_10_10 011_10_01_10_10_10_10_01
100_01_01_10_01_01_01_10 011_10_10_01_01_10_10_10
100_01_01_01_10_10_01_01 011_01_10_10_10_01_10_10
100_100_011_100_01_10_01_01 100_011_100_100_01_01_01_10
100_011_011_011_10_01_10_10 011_011_100_011_10_10_01_10
100_100_011_100_10_01_01_01 100_011_011_011_10_10_10_01
100_011_100_100_10_01_01_01 011_100_011_100_01_01_10_10
100_011_011_011_10_10_01_10 011_011_011_100_01_10_10_10
100_100_011_011_01_01_10_10 011_011_011_100_10_01_10_10
011_100_011_100_10_10_01_01 100_100_100_10_10_01_01_01
011_011_011_10_10_01_01_10 100_100_100_10_01_10_01_01
100_100_100_01_01_01_10_10
```

```
100_10_01_10_01_01_01_01 100_01_01_10_01_10_01_01
100_01_01_01_10_01_10_01 100_01_01_01_10_01_01_10
100_01_10_01_01_01_10_01 011_10_10_10_10_01_01_10
100_01ī_1̄00_1̄00_01_01_10_01 01̄1_01ī_1̄00_01ī_10_01_10_10
011_100_100_100_01_10_01_01 011_100_100_100_10_01_01_01
011_100_011_100_10_01_01_10 011_011_011_100_10_10_01_10
011_011_100_100_10_10_01_01 011_011_011_01_0̄1_1̄0_1̄0_1̄0
011_011_011_01_1̄0_1̄0_1̄0_01 100_1̄00_1̄00_01ī_01_01_10_01
011_011_011_10_01_10_01_10 011_011_011_100_10_10_10_01
100_100_100_01_01_10_01_10 011_011_011_10_1̄0_01_1̄0_01
100_100_100_10_01_01_10_01
Total number of SAX nodes without a match: 46
```

As we are using the same promotion strategy for both iSAX indexes, the output shows that the two iSAX indexes have a different structure, hence the dissimilarities in the list of SAX representations without a match on the second iSAX index. Additionally, we can see that the maximum cardinality in the printed SAX representations is just 8 and that most SAX words have a cardinality of 4, which means that there were not so many splits.

Lastly, keep in mind that, in general, the smaller the number of segments in a SAX representation, the smaller the number of nodes without a match is going to be. Additionally, the larger the threshold value, the smaller the number of nodes without a match is going to be, because large threshold values minimize splits. In general, *the smaller the number of possible SAX representations is, the smaller the number of nodes without a match is going to be*.

This can be seen in the following output, where we have reduced the number of segments to 4 and increased the cardinality to 64:

```
$ ./noMatch.py -s 4 -c 64 -t 1500 -w 128 500k.gz 506k.gz
Max Cardinality: 64 Segments: 4 Sliding Window: 128 Threshold: 1500
Default Promotion: False
101_01_01_10 010_10_01_10 1001_0110_0110_100 01100_1001_1000_0111
01100_01ī1_1000_ī001 01ī00_1001_01ī1_1000 10011_1000_01ī0_01ī1
01100_1000_0111_1001 011110_01ī1_1̄0000_10000_01̄01_1000_1000_100
1000_0101_1̄000_ī00 0111_01ī1_011_ī01 01ī11_01101_1̄001_1̄000
1̄0000_01ī01_10000_10001 1̄0001_01ī10_10001_01̄110 01ī10_01101_1001_1000
01ī10_10001_01110_10001 01111_10010_01ī10_01111 01ī10_1̄0000_01ī10_1001
01101_10001_1̄0000_10000 10001_10010_0111_01̄10 01101_1̄0000_1̄0001_1̄0000
01110_01ī10_10010_1000

1001_0110_1001_011 10011_0110_1000_0111 011110_10000_01111_10000
1001̄0_01ī1_01ī1_01110 1̄0001_01ī0_01110_1000ī
10001_10010_0110_01̄11 10001_1̄0000_01̄101_1̄0000 10010_01111_01110_01111
01110_01111_011ī1_10010 01ī1_01ī0_01ī1_10010 10001_10001_01ī01_0111
Total number of SAX nodes without a match: 34
```

In this case, we have fewer terminal nodes without a match.

The next section briefly touches on the topic of testing Python code by writing three basic Python tests for the isax package.

Writing Python tests

In this last section of this chapter, we are going to learn about Python testing and write three tests for our code with the help of the `pytest` package.

As the `pytest` package is not installed by default, the first task you should carry out is installing it using your favorite method. Part of the `pytest` package is the `pytest` command-line utility, which is used for running the tests.

> **Unit testing**
>
> In this section, we are writing unit tests, which are usually functions that we write to make sure that our code works as expected. The result of a unit test is either PASS or FAIL. The more extensive the unit testing is, the more useful it is.

After a successful installation, if you execute the `pytest` command on a directory that does not contain any valid tests, you are going to get information about your system and your Python installation. On a macOS machine, the output is the following:

```
$ pytest
========================== test session starts ==========
platform darwin -- Python 3.10.9, pytest-7.2.1, pluggy-1.0.0
rootdir: /Users/mtsouk/code/ch05
collected 0 items

================ no tests ran in 0.00s =====================
```

As far as testing functions are concerned, there is a simple rule that you have to keep in mind. A testing function when using the `pytest` package is any Python function that is prefixed with `test_` in a file where its filename is also prefixed by `test_`.

There are many tests that could be written. Usually, we want to test as much as possible, starting from the most critical parts of the code and moving to the less critical ones. For the purposes of this chapter, we have decided to test the core logic of the implementation by writing three tests.

The next subsection talks about the three tests that we are going to implement in this chapter in more detail.

What are we going to test?

The first thing to define is what we are going to test and why. For this chapter, we are going to write the following three tests:

- We are going to count the number of subsequences in an iSAX index and make sure that the iSAX index holds all the subsequences.

- We are going to test the number of node splits of an iSAX construction – this time, the correct number of splits is going to be stored in a global variable.

- Lastly, we are going to join the same time series with itself. This means that we should get a list of Euclidean distances where all values are equal to 0. Keep in mind that as we are talking about floating-point numbers, the Euclidean distances might be very close to 0 but not exactly 0.

The list of tests is far from complete, but it is a good way to illustrate the use and usefulness of tests.

The filenames of the time series and the iSAX parameters, as well as the number of splits and subsequences, are going to be given as global variables in the source code file that holds the testing code for reasons of simplicity. If you want to dynamically pass parameters to `pytest` tests, visit the *Basic patterns and examples of pytest* link in the *Useful links* section at the end of the chapter for more information.

Comparing the number of subsequences

In this test, we compare the number of subsequences in an iSAX index to the theoretical number of subsequences based on the sliding window size and the time series length.

The relevant code is as follows:

```
def test_count_subsequences():
    variables.nSplits = 0
    variables.segments = segments
    variables.maximumCardinality = cardinality
    variables.slidingWindowSize = slidingWindow
    variables.threshold = threshold
    i, ts = createISAX(TS, slidingWindow, segments)

    sum = 0
    for k in i.ht:
        t = i.ht[k]
        if t.terminalNode:
            sum += t.nTimeSeries()

    assert sum == len(ts) - slidingWindow + 1
```

First, we appropriately set the global variables in `./isax/variables.py` based on the global values found in the preamble of `test_isax.py`.

The `createISAX()` helper function is used to create iSAX indexes for testing. You have seen that function before in the `speed.py` utility.

What is important and closely related to the test is the use of the `assert` keyword. `assert` checks the trueness of the statement that follows. If the statement is `True`, then the `assert` statement passes. Otherwise, it throws an exception, and as a result, the test function fails. The `assert` keyword is used in all our test functions.

Next, we are going to discuss the test that checks the number of node splits.

Checking the number of node splits

For the purposes of this test, we assume that we have a different program in any programming language that we consider correct that gives us the actual number of node splits. This number of node splits is stored in a global variable (`splits`) and read by the test function.

The relevant Python code is as follows:

```
def test_count_splits():
    variables.nSplits = 0
    variables.segments = segments
    variables.maximumCardinality = cardinality
    variables.slidingWindowSize = slidingWindow
    variables.threshold = threshold
    variables.defaultPromotion = False
    i, ts = createISAX(TS, slidingWindow, segments)

    assert variables.nSplits == splits
```

In this code, we first appropriately set the global variables in `./isax/variables.py` based on the global values found in `test_isax.py`. Do not forget to reset `variables.nSplits` and select the correct promotion strategy (`variables.defaultPromotion`) in your test function.

The last test function computes the join of a time series with itself, which means that all Euclidean distances after the join should be equal to `0`.

All Euclidean distances are 0

In this test, we are going to create the iSAX index of a time series and join it with itself. As we are comparing a time series with itself, the list of Euclidean distances should only *contain zeros*. Therefore, with this unit test, we examine the logical correctness of our code.

The relevant Python test function is implemented as follows:

```
def test_join_same():
    variables.nSplits = 0
    variables.segments = segments
    variables.maximumCardinality = cardinality
    variables.slidingWindowSize = slidingWindow
```

```
        variables.threshold = threshold
        i, _ = createISAX(TS, slidingWindow, segments)
        Join(i, i)

        assert np.allclose(variables.ED, np.zeros(len(variables.ED))) ==
True
```

First, we appropriately set the global variables in ./isax/variables.py based on the global values found in test_isax.py. After that, we call createISAX() to construct the iSAX index and then call the Join() function to populate the list of Euclidean distances.

The NumPy zeros() function creates a NumPy array with all zeros. Its parameter defines the length of the NumPy array that is going to be returned. The NumPy allclose() function returns True if its two NumPy array arguments are equal within a tolerance. This is mainly the case because when using floating-point values, there might be small differences due to rounding.

In the next subsection, we are going to run the tests and see the results.

Running the tests

In this section, we are going to run the tests and see their results. All the previous code can be found in the ./ch05/test_isax.py file.

The results of the tests, which are all successful, are as follows:

```
$ pytest
================== test session starts =====================
platform darwin -- Python 3.10.9, pytest-7.2.1, pluggy-1.0.0
rootdir: /Users/mtsouk/TSi/code/ch05
collected 3 items

test_isax.py ...                                    [100%]

============== 3 passed in 2784.53s (0:46:24) ===============
```

In the case that there were one or more failed tests, the output would look like the following (in this case, only one test failed):

```
$ pytest
======================== test session starts =================
platform darwin -- Python 3.10.9, pytest-7.2.1, pluggy-1.0.0
rootdir: /Users/mtsouk/TSi/code/ch05
collected 3 items

test_isax.py .F.                                    [100%]
```

```
=====================FAILURES ============================
_____ test_count_splits _____

    variables.nSplits = 0
    variables.segments = segments
    variables.maximumCardinality = cardinality
    variables.slidingWindowSize = slidingWindow
    variables.threshold = threshold
    _, _ = createISAX(TS, slidingWindow, segments)

>   assert variables.nSplits == splits
E   assert 5669 == 5983
E    +  where 5669 = variables.nSplits

test_isax.py:58: AssertionError
================ short test summary info ==================
FAILED test_isax.py::test_count_splits - assert 5669 == 5983
=========== 1 failed, 2 passed in 2819.21s (0:46:59) ======
```

The good thing is that the output shows the reason that one or more tests have failed and includes the relevant code. In this case, it is the `assert variables.nSplits == splits` statement that failed.

This is the last section of this chapter, yet the most important one, as testing can save you lots of time during development. The main purpose of our tests is to test the logic and the correctness of the code, which is very important.

Summary

In this chapter, we have seen code for testing the speed of an iSAX index and joining two iSAX indexes based on the SAX representations of their nodes. Then, we briefly discussed the subject of testing Python code and implemented three tests for the `isax` package.

We also discussed the joining of iSAX indexes, which is based on the node types. Additionally, the tests we carried out made sure that the core logic of our code is correct.

In the next chapter, we are going to learn how to visualize iSAX indexes to better understand their structure and performance.

Before you start reading and working through *Chapter 6*, experiment with the command-line utilities that we have developed in this chapter and try to create your own.

Useful links

- The pytest package: `https://pypi.org/project/pytest/`
- Official pytest documentation: `https://docs.pytest.org/`
- *Practices of the Python Pro*, written by Dane Hillard
- *Mastering Python, 2nd Edition*, written by Rick van Hattem
- *Robust Python: Write Clean and Maintainable Code*, written by Patrick Viafore
- *Python Testing with pytest, 2nd Edition*, written by Brian Okken
- Basic patterns and examples of pytest: `https://docs.pytest.org/en/latest/example/simple.html`

Exercises

Try to work through the following exercises:

- Use `accessSplit.py` to learn how the sliding window size affects the construction speed of the `2M.gz` time series from *Chapter 4*. Perform your experiments for the following sliding window sizes: `16`, `256`, `1024`, `4096`, `16384`, and `32786`.

- Can you resolve the overflow situations with `accessSplit.py` and the `500.gz` time series we came across at the beginning of the chapter?

- Try reducing the threshold values in the `speed.py` examples presented in the *Checking the search speed of iSAX indexes* section and see what happens.

- Create two time series with 250,000 elements each and use `speed.py` to understand their behavior when the number of segments is in the 20 to 40 range. Do not forget to use an appropriate sliding window size.

- Experiment with `speed.py` but this time, change the threshold value instead of the number of segments. Is the threshold value more important than the number of segments in the search speed of an iSAX index?

- Modify `speed.py` to display the number of misses in subsequence queries.

- Modify `join.py` to print the time it took to perform the join.

- Modify `saveLoadList.py` to include the threshold value in the filename where we save the contents of the list with the Euclidean distances.

- Run the `pytest` command on your own machines and see the output that you get.

6
Visualizing iSAX Indexes

In the previous chapter, we learned about comparing and joining iSAX indexes. However, it is still difficult to imagine the structure and the height of an iSAX index without seeing it as an image.

And although some people prefer text, some other people prefer log files, and some others prefer numbers, almost all people like good-looking and informative visualizations. Additionally, all people understand the importance of having a high-level view of their data. This includes iSAX indexes and tree structures in general, mainly because there is no other practical way to perform the same task, especially when working with big time series.

In *Chapter 1*, we saw how to visualize a time series. This chapter is all about visualizing iSAX indexes in order to get a better understanding of their size, shape, and structure.

Visualizing large structures and trees such as iSAX indexes is not a trivial process but of a trial-and-error one. As no single visualization can do the job, we are going to try different kinds of plots and see what they tell us about the iSAX index. Therefore, you should expect to see lots of visualizations in this chapter, and I expect that you are going to create many more visualizations on your own while reading this book.

In this chapter, we are going to cover the following main topics:

- Storing an iSAX index in JSON format
- Visualizing an iSAX index
- Trying something radical
- More iSAX index visualizations
- Using icicle plots
- Visualizing iSAX as a Collapsible Tree

Technical requirements

The GitHub repository for the book is at `https://github.com/PacktPublishing/Time-Series-Indexing`. The code for each chapter is in its own directory. Therefore, the code for *Chapter 6* can be found in the `ch06` folder. However, in this chapter, there exist many directories under the `ch06` folder that contain the code for the different visualizations that we are going to create – this is a good way to organize code.

Storing an iSAX index in JSON format

For the visualizations of this chapter, we are going to use the low-level D3.js JavaScript library.

> **Is D3.js the only way to create visualizations?**
>
> The powerful D3.js JavaScript library is not a panacea and therefore, it is not the only way to create visualizations. There exist many Python packages that are good at plotting data, as well as programming languages such as R or Julia. However, JavaScript can be used for presenting your plots in a web page environment, which is not usually the case with the other options.

For the JavaScript D3.js code to work, we need to represent an iSAX index in **JSON** format so that it can be understood by the JavaScript code – we mainly need to represent **the structure and the connections** between iSAX nodes in a way that can be understood by a computer and a programming language. Therefore, the first step we should take is to convert an iSAX index representation with its structures from Python code into a different structure made by JSON records.

Although this JSON format is not universal and might fail in some cases, it is going to be used throughout this chapter as all presented D3.js code works fine with it – all presented examples are tested and fully working.

First, we need to visit `https://d3js.org/` and click on **Examples** at the top of the page, which is going to take us to `https://observablehq.com/@d3/gallery`. The latter page is going to bring us to a page with professional, functional, and beautiful plots that look appropriate for the kind and amount of data that we want to plot in this chapter.

From the long list of available visualizations, we need to pick the ones that we prefer and are a good match for our data and its structure – our first attempt might not be perfect. Do not forget that iSAX indexes can have a large number of nodes. Therefore, we should think rationally and pick something that is going to look good with lots of data.

From that list, we pick `Tree, Tidy`. Behind the visualization, there is JavaScript code embedded into HTML that reads the JSON data, parses it, and creates the visualization.

Now that we have found our visualization of preference (`https://observablehq.com/@d3/tree`), we might begin looking at the JavaScript code to get a better idea of the data format that is expected from the JavaScript code. However, what is more important is the JSON record format.

> **Where is the JavaScript code?**
>
> JavaScript is a powerful but low-level programming language. The good thing is that the presented visualizations do not need any JavaScript knowledge to work. You just need to put your own data in the right format, at the right place, and that is all!

The JSON records that we are going to support should have the following format – this format was found by looking into the JSON file that the JavaScript code uses to get its data:

```
"name": "flare",
"children": [
  {
    "name": "analytics",
    "children": [
      {
  .
  .
  .
```

The common idea behind the structure that we want to support is that we have a root node – the root of the tree – that has multiple children, those children have children of their own, and so on. The name of each node is a **string** value, and the children of each node are kept in an **array**. This is the most important information about the JSON file format. Additionally, the first value of the name field is given to the root node of the tree – in this case, that name will be `flare`.

There is an additional field that is going to be needed in some of the presented visualizations. Later on in this chapter, we are going to learn that terminal nodes have an additional field for storing the number of the subsequences that they hold – this is not used by every visualization. However, the thought remains the same.

A sample tidy tree visualization with custom data can be seen in the following figure. We are going to present more complex visualizations in the sections that follow. This is a simple tree structure with a root node and 13 nodes as its children:

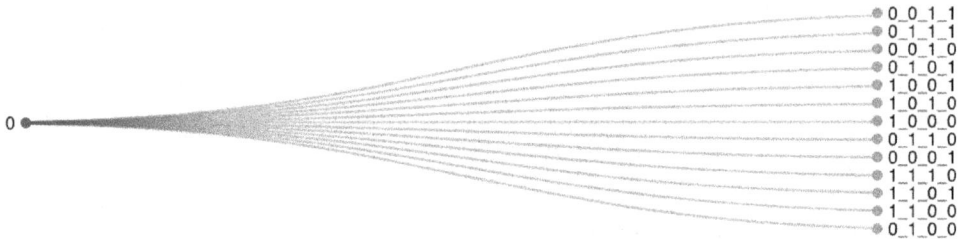

Figure 6.1 – A tree visualization

With all that in mind, we can now begin working on our Python code. The Python script for representing an iSAX index as a JSON file is called exportJSON.py. The logic behind exportJSON.py is that after creating the iSAX index, we traverse it in order to generate the JSON output in the desired format.

But first, here is the definition of the JSON record that is going to be used in the Python script:

```
JSON_message = {
    "name": None,
    "size": None,
    "children": []
}
```

This is the basic JSON record format – we can add more fields to this record depending on our needs for any customization requirements without breaking any JavaScript code, as JavaScript is going to read only the fields that it needs. Although the format is defined in the code, it is not used by the Python code, mainly because Python does not need a predefined structure for JSON data that is stored in dictionaries. However, it is good to have it defined as a point of reference. The size field stores the total number of subsequences stored under each terminal node. Inner nodes do not need such a field.

The Python code within exportJSON.py that reads the existing iSAX index and prints the JSON output can be found at the end of the main() function:

```
# The JSON data to return
data = {}
data['name'] = "0"
data['children'] = []

# Create JSON output
for subTree in ISAX.children:
    if ISAX.ht[subTree] == None:
        continue
```

```
        subTreeData = createJSON(ISAX.ht[subTree])
        data['children'].append(subTreeData)

    print(json.dumps(data))
```

The name of the Python dictionary that holds the JSON records is `data`. By default, the root node has the name 0, which is a string – this is analogous to `flare`. You can change that to anything you like.

The previous code visits and processes the children of the root node of the iSAX index only. The rest is handled by the `createJSON()` function. `createJSON()` is the function that actually generates the JSON output by adding data for the current subtree that is being examined. The `data` variable holds all JSON data.

The last statement prints all JSON records on the screen using `json.dumps()`.

The `createJSON()` function is implemented as follows:

```
def createJSON(subtree):
    if subtree == None:
        return None

    t = {}
    t['name'] = subtree.word
    t['children'] = []

    # First, check if this is a Terminal node
    if subtree.terminalNode == True:
        t['size'] = subtree.nTimeSeries()
        return t
    # This is still a Terminal node
    # Just in case!
    elif subtree.left == None and subtree.right == None:
        print("This should not happen!")
        return t
    else:
        ch1 = createJSON(subtree.left)
        ch2 = createJSON(subtree.right)
        t['children'].append(ch1)
        t['children'].append(ch2)

    return t
```

The `createJSON()` function is called recursively in order to visit all the nodes of each subtree. This mainly happens because we need to process all inner nodes and all terminal nodes.

> **About the JSON output**
>
> This particular Python script generates JSON output based on a particular JSON record format. Once you get the idea right, it is going to be easy to make small changes to the script, add more fields to the JSON records, or create something totally different. All these depend on the format that the visualization script expects to work with.

For terminal nodes, we keep the number of the subsequences they hold – this takes place in the `size` field, and the processing of each terminal node ends here. However, for inner nodes, we recursively call `createJSON()` in order to process the left and the right children or subtrees of each inner node.

With all that in mind, let us see `exportJSON.py` in action. First, we are going to use it with a small iSAX index, using a time series named `ts.gz` that contains 100 elements – due to its small size, `ts.gz` is going to be used for experimentation purposes. `ts.gz` was created by running `../ch01/synthetic_data.py 100 -10 10`, saving the output in a file named `ts` and compressing `ts` with `gzip(1)`.

Running `exportJSON.py` with `ts.gz` produces the following kind of output:

```
$ ./exportJSON.py -s 3 -c 8 ts.gz
{"name": "0", "children": [{"name": "0_0_1", "children": [], "size":
18}, {"name": "0_1_1", "children": [], "size": 12}, {"name": "1_0_1",
"children": [], "size": 12}, {"name": "1_0_0", "children": [], "size":
17}, {"name": "0_1_0", "children": [], "size": 12}, {"name": "1_1_0",
"children": [], "size": 13}, {"name": "1_1_1", "children": [], "size":
1}]}
```

Processing that data with the `jq(1)` utility, which beautifies JSON records, generates the next better-looking output – in this case, all the children of the root are terminal nodes:

```
$ ./exportJSON.py -s 3 -c 8 ts.gz | jq
{
  "name": "0",
  "children": [
    {
      "name": "0_0_1",
      "children": [],
      "size": 18
    },
    {
      "name": "0_1_1",
      "children": [],
      "size": 12
    },
    {
      "name": "1_0_1",
```

```
        "children": [],
        "size": 12
    },
    {
        "name": "1_0_0",
        "children": [],
        "size": 17
    },
    {
        "name": "0_1_0",
        "children": [],
        "size": 12
    },
    {
        "name": "1_1_0",
        "children": [],
        "size": 13
    },
    {
        "name": "1_1_1",
        "children": [],
        "size": 1
    }
    ]
}
```

Bear in mind that the output depends on the parameters of the iSAX index. Different iSAX parameters generate different outputs and different tree structures.

Now, let us try exportJSON.py with a bigger time series that is named 100k.gz, which contains 100,000 elements and was created as follows:

```
$ ../ch01/synthetic_data.py 100000 -100 100 > 100k
$ gzip 100k
```

We ran exportJSON.py with 100k.gz as follows:

```
$ ./exportJSON.py -s 4 -c 16 -t 2500 100k.gz > 100k.json
```

The output file was saved as 100k.json. The reason for using a threshold value of 2500 is to have a more compact tree. However, at the end of the day, what matters is your needs and the actual parameters of your iSAX indexes.

At this point, we processed `ts.gz` and `100k.gz` with `exportJSON.py`, and we end up having two JSON files named `ts.json` and `100k.json`, respectively. Although we might need `ts.json` for testing, all coming visualizations are going to use `100k.json`.

The next subsection is about downloading the JavaScript project on our local machine and executing it from there.

Downloading the JavaScript code locally

Observablehq allows you to download the code locally, make changes to it, and change the contents of the JSON file that gets visualized – as I do not know JavaScript, I find that this is the easiest way to use `D3.js` and its powerful capabilities. Therefore, in this subsection, we are going to learn how to do so.

In every visualization from `https://observablehq.com/@d3/gallery`, there is a menu that appears when we click on the three dots that appear near the upper-right corner of the web page. From that menu, click on the **Export** link, which displays a submenu. From that submenu, we should click on the **Download code** option. This is going to download the contents of the current project on our local machine as a compressed file that we should extract and use.

For the purpose of this section, we are going to download the JavaScript project found at `https://observablehq.com/@d3/tree`, which is going to download a file named `tree.tgz`. After we have uncompressed that file, we are going to get a directory called `tree`. It is not necessary to fully understand the contents of the directory, but it helps. However, you need to know the path to the JSON file with the data.

The output of the `tree(1)` utility, which lists the contents of directories in a tree-like format, when examining the contents of the `tree` directory, is the following:

```
$ tree
.
├── 5432439324f2c616@268.js
├── 7a9e12f9fb3d8e06@498.js
├── LICENSE.txt
├── README.md
├── files
│   └── 85b8f86120ba5c8012f55b82fb5af4fcc9ff5e3cf250d110e111b3ab
98c32a3fa8f5c19f956e096fbf550c47d6895783a4edf72a9c474bef5782f
879573750ba.json
├── index.html
├── index.js
├── inspector.css
├── package.json
└── runtime.js

2 directories, 10 files
```

The JSON file with the records is located in the `files` directory – this is the file that we need to overwrite with our own data file. In order to load the project, we need to access the `index.html` file, which is going to load all the necessary dependencies.

We only need to put our own JSON data into the `files` folder – this is the only required change to be done.

The next subsection is about running the downloaded JavaScript project on your own machine, which requires running your own local HTTP server.

Running the code locally

The process of running the code locally includes the following steps:

1. Go into the directory with the code.
2. Make changes to the JSON file with the data – each example has its own JSON data file in a directory named `files`.
3. Run a local HTTP server.

The JSON file with the data has a long and strange filename that is embedded into the JavaScript code. I am not so proficient with JavaScript, so I am going to use the default filename, which is located in the `files` directory. For security reasons, it is not allowed for the web server to access files outside of its root directory. Therefore, we need to *copy the JSON file* we have created with `exportJSON.py` in the `files` directory of each individual project and *overwrite the existing JSON file*, even if we are using the same one in all our examples.

The next subsection shows how to run your own local HTTP server and view the JavaScript code in action.

Running a local HTTP server

The easiest way to run a local HTTP server is by executing `python3 -m http.server` in the directory that interests you. If everything goes fine, the HTTP server is going to listen to port number `8000` and it is going to be accessed as `http://localhost:8000/`. This is much easier than it looks.

This process is going to be used throughout this chapter. All the required documents and files are in the GitHub repository of the book, so you have nothing more to download. If you want to experiment, just change the contents of the JSON file with the data.

The next subsection shows how to test the process.

Testing the process

I have renamed the directory of the previous project from `tree` to `TreeTidy` – it is a good practice to use descriptive directory names.

So, first, we need to go to the `ch06` directory of the GitHub repository for this book and then go to the `TreeTidy` directory. After that, we need to run `python3 -m http.server`. Now, we have an HTTP server running on our local machine that listens to the `8000` TCP port. Therefore, we need to point our web browser to `http://localhost:8000/` and see the generated visualization.

The generated output from the Python web server is going to look as follows:

```
$ python3 -m http.server
Serving HTTP on :: port 8000 (http://[::]:8000/) ...
::ffff:127.0.0.1 - - [29/Mar/2023 20:42:08] "GET / HTTP/1.1" 200 -
::ffff:127.0.0.1 - - [29/Mar/2023 20:42:08] "GET /inspector.css
HTTP/1.1" 200 -
::ffff:127.0.0.1 - - [29/Mar/2023 20:42:08] "GET /runtime.js HTTP/1.1"
200 -
::ffff:127.0.0.1 - - [29/Mar/2023 20:42:08] "GET /index.js HTTP/1.1"
200 -
::ffff:127.0.0.1 - - [29/Mar/2023 20:42:08] "GET
/5432439324f2c616@268.js HTTP/1.1" 200 -
::ffff:127.0.0.1 - - [29/Mar/2023 20:42:08] "GET
/7a9e12f9fb3d8e06@498.js HTTP/1.1" 200 -
::ffff:127.0.0.1 - - [29/Mar/2023 20:42:08] code 404, message File not
found
::ffff:127.0.0.1 - - [29/Mar/2023 20:42:08] "GET /favicon.ico
HTTP/1.1" 404 -
```

If you see any error messages in the generated output, you should try to resolve them.

However, unless you are in the wrong directory or there is another TCP service running on TCP port `8000`, there should be no issues.

So far, we have learned how to represent an iSAX index in JSON format and how to download the JavaScript projects from `https://observablehq.com/`.

In the section that follows, we will begin our iSAX visualization journey.

Visualizing an iSAX index

In this section, we are going to begin visualizing iSAX indexes.

As in most areas of computing, your visualizations are going to improve over time. The first visualizations are usually less beautiful and/or informative than later ones. So, we are going to experiment and try things before we end up with a good-looking iSAX visualization.

As visualizations include personal taste, your visualization of choice might differ from the ones used in this chapter. However, we need to start doing and improve in the process!

Let us begin with the visualization of the next subsection.

A personal story

At the time of writing this book, I am doing research related to iSAX. In one of my experiments, I ran a utility that creates two iSAX indexes and joins them in a more sophisticated way than the one presented in *Chapter 5*. The utility processed 2 time series with 500,000 elements each and ran for more than 18 days! Additionally, it took the same utility about 2 hours to process 2 time series with 1,500,000 elements each, which means that the utility works well. I decided to visualize each iSAX index using a separate Python utility. Long story short, in the case of the time series with 500,000 elements, I mistakenly used 32 segments and a cardinality value of 4, instead of 4 segments and a cardinality value of 32! This means that the root of each iSAX index had 2^{32} children! Therefore, joining them included so many calculations, which explained the fact that the utility still ran after 18 days. If I had visualized each iSAX index earlier, I would have found the issue much sooner.

Visualizing iSAX as a tree

In this first try, we are going to visualize an iSAX index as a tree using various visualizations. As iSAX has a tree structure, using this kind of visualization makes perfect sense.

For this subsection, we are going to use the visualization stored in the `TreeTidy` directory that we saw earlier. The first task to up is to update the JSON file stored in the `files` directory of the `TreeTidy` directory. If we are in the `TreeTidy` directory, we can run `cp ../100k.json files/85b8f8…9573750ba.json`. The full filename is omitted for brevity – just make sure that you use the correct filename with the help of `shell` auto-completion.

In *Figure 6.2*, you can see the visualization of the iSAX index generated for the `100k.gz` time series using the `D3.js` code that also generated the sample output of *Figure 6.1*.

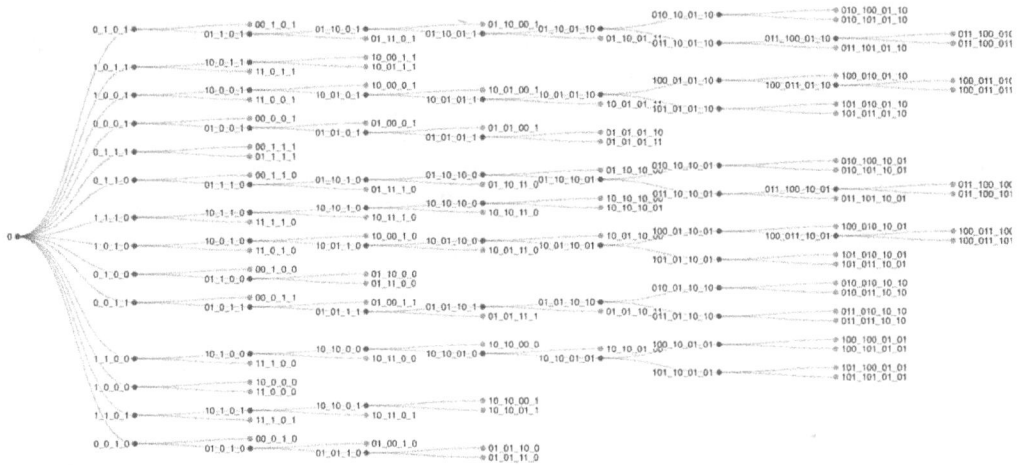

Figure 6.2 – Visualizing an iSAX index as a tree

What does *Figure 6.2* tell us? It tells us that we are dealing with a relatively small iSAX index that is pretty balanced (the depths of the terminal nodes do not differ too much), which is a good thing. By default, terminal nodes are visualized with gray color circles, whereas inner nodes are black.

So, what can we do next? Next, we can try visualizing `100k.gz` using different iSAX parameters. So, in this case, we are going to use the following parameters:

```
$ ./exportJSON.py -s 4 -c 16 -t 5000 100k.gz
```

As before, the generated output is going to be stored in the existing JSON file inside the `files` directory. The updated output can be seen in *Figure 6.3*:

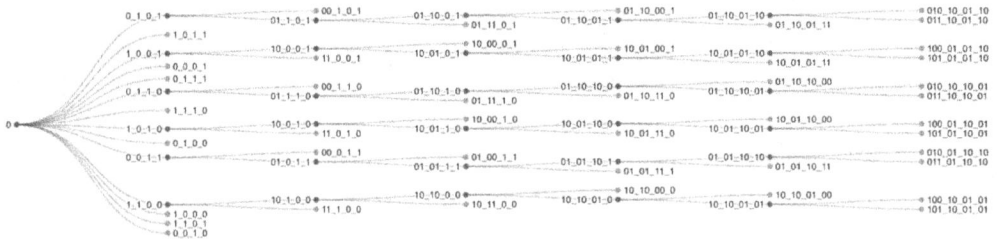

Figure 6.3 – Increasing the threshold value to 5,000

As expected, the iSAX index is smaller than before, mainly because we have fewer node splits. However, it looks less balanced than the iSAX presented in *Figure 6.2*.

Is there anything more to do? We can experiment a little bit more and change the segments value from 4 to 3 while keeping the threshold value at 5000.

So, this time, the JSON output is going to be generated using this command:

```
$ ./exportJSON.py -s 3 -c 16 -t 5000 100k.gz
```

After that, we need to store the output in the JSON file located in the `files` directory. The new visualization can be seen in *Figure 6.4*:

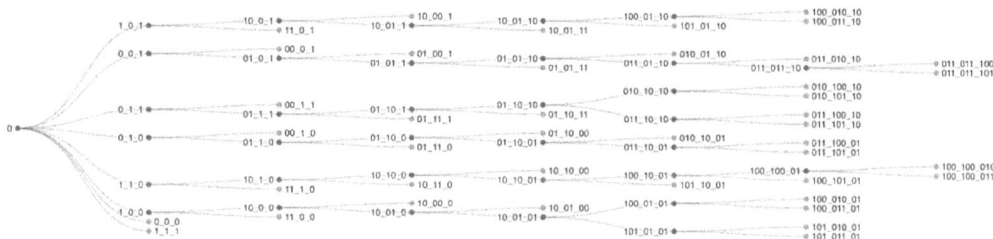

Figure 6.4 – Visualizing an iSAX with three segments

As expected, the root node has fewer children. However, the general shape of the iSAX index presented in *Figure 6.4* is similar to the one presented in *Figure 6.3*. In my personal opinion, *Figure 6.2* shows a better and more balanced iSAX index compared to the other two versions. Balanced trees, and therefore balanced iSAX indexes, are generally faster to search, which is a desired property.

In this section, we saw how to visualize iSAX indexes as tree structures, which makes perfect sense, as iSAX indexes are trees.

In the next section, we are going to try a different kind of visualization for the iSAX index structure. After all, visualization and experimentation are good friends.

Trying something radical

In this section, we are going to try a different kind of visualization for visualizing an iSAX index, just in case it reveals any extra kind of information. So, we are going to use a **Radial Tree** structure instead of a regular tree. For that, we need to go to the `TreeRadialTidy` directory inside the `ch06` directory and replace the JSON file found in the `files` directory with `100k.json` – the correct file is already there. However, if you want to use your own data, you should update that file.

Next, we should run the Python HTTP server and point our web browser to `http://localhost:8000/`. The generated output is presented in *Figure 6.5*:

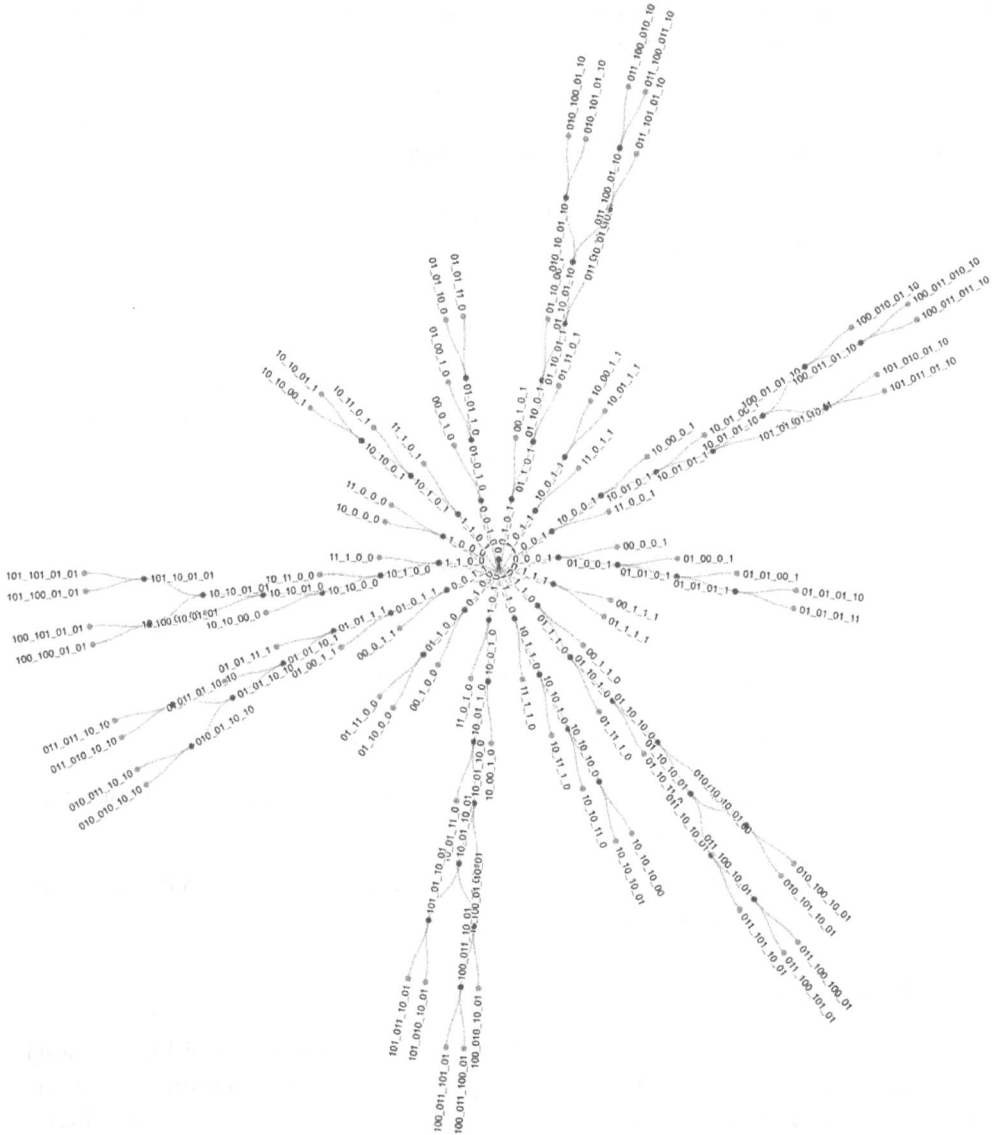

Figure 6.5 – Using a Radial Tree structure

What kind of information can we get from *Figure 6.5*? Is that better than a regular tree structure? I do not know whether it is better or not, but it surely presents the same information in a totally new way!

One advantage of the radial tree is that it performs better when dealing with iSAX indexes with large depths as they can fit better on screen. Personally, I believe that the plain tree structure is more suitable for iSAX indexes than the radial tree.

The next section continues the visualization process of iSAX indexes by trying more visualizations.

More iSAX index visualizations

We are not done yet! There exist ways to improve the previous visualization by adding more information to the output as well as the ability to compress various parts of it – there is always the danger of *putting too much information* on a graph or a plot, but we are not going to make that mistake here.

First, we are going to go to the `ZoomableTreemap` directory in order to try a zoomable structure named **Zoomable Treemap**, which is better when dealing with large iSAX indexes.

The Zoomable Treemap uses an additional attribute called `value`. In this case, I had two choices: either change the output of the Python script or change the JavaScript code. I decided to do the latter. So, I changed the `value` attribute in the JavaScript code to `size`, which is what the Python script generates. However, in our case, this created a bug in the JavaScript code related to the sum of the presented values, which means that this was not the correct decision.

Therefore, we are going to change the JSON file and replace the `size` field name with `value`.

As before, we should overwrite the JSON file in the `files` directory with `100k.json` and run the Python HTTP server. The generated output can be seen in *Figure 6.6*:

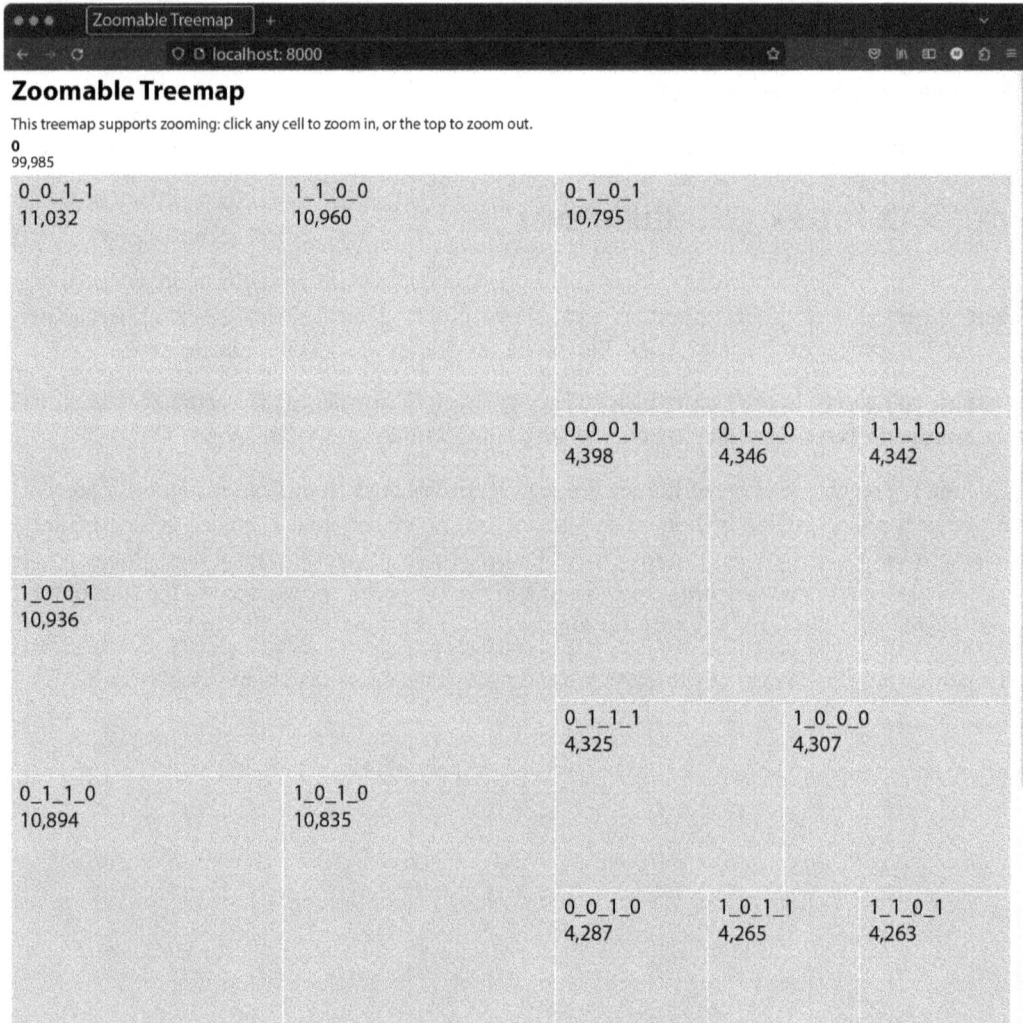

Figure 6.6 – A Zoomable Treemap visualization

It turns out that the Zoomable Treemap might be difficult to read and understand – it is even difficult to realize that we are talking about a tree structure. Therefore, it might not be a good choice for the iSAX visualization. However, the *zoomable* capability is very handy in almost all cases.

If we used the buggy version, then instead of the numeric values in the output, we would have got a NaN value – this probably had to do with the JavaScript code.

Let us now continue with something different. Go to the ZoomableSunburst directory and replace the file in the files directory with the 100k.json file. Once again, we need to make code changes. Specifically, we need to replace the value field used in 86ddbc29bd33f9d6@357.js with the size field that our JSON record has. The code stored in GitHub has all the necessary changes in it. The generated output can be seen in *Figure 6.7*:

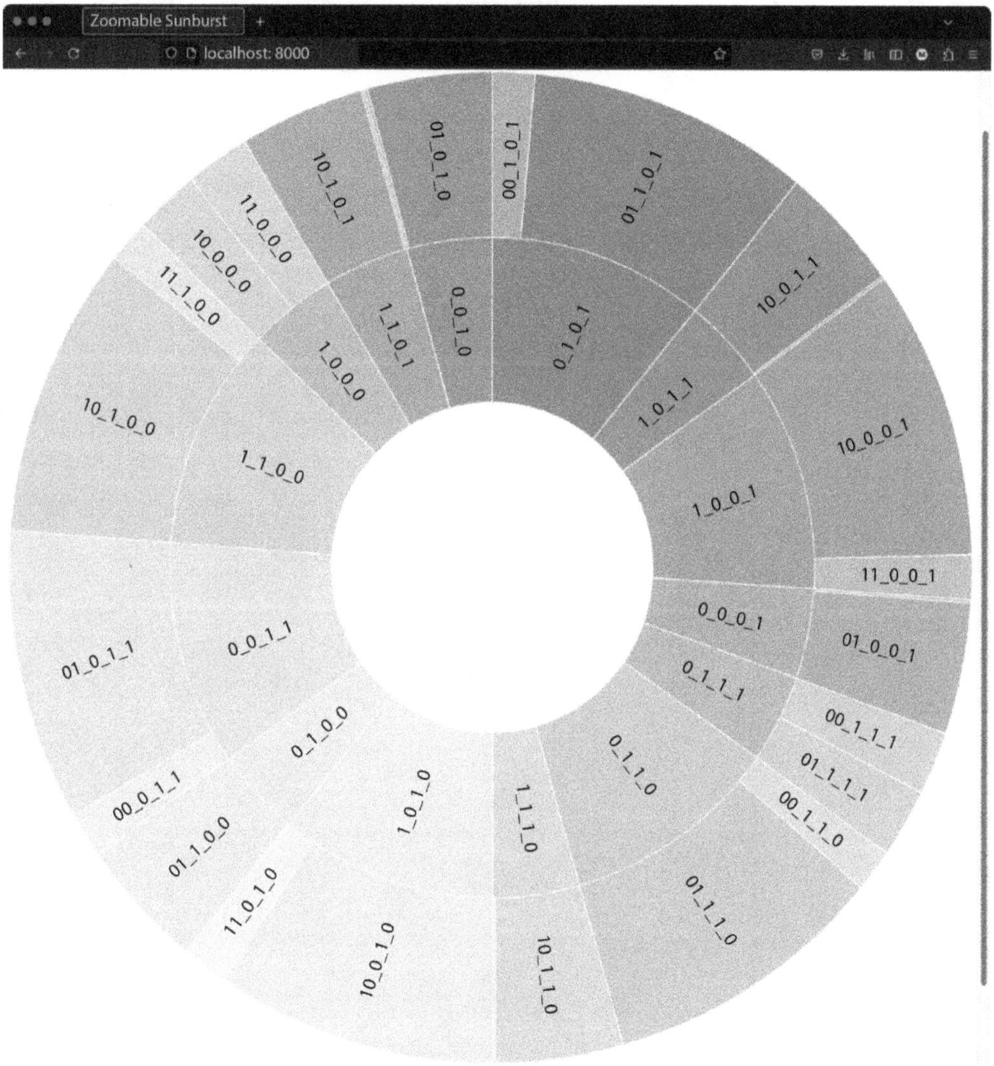

Figure 6.7 – A Zoomable Sunburst visualization

The main advantage of this visualization is that it does not display the entire iSAX from the beginning, but it can do so as we are zooming in on the visualization by clicking on the different parts of the sunburst. So, it hides some information, which can be displayed on demand.

If we zoom into any part of the Sunburst, we are going to get a closer look at that particular part of the iSAX index.

Figure 6.8 shows such a part of the Sunburst:

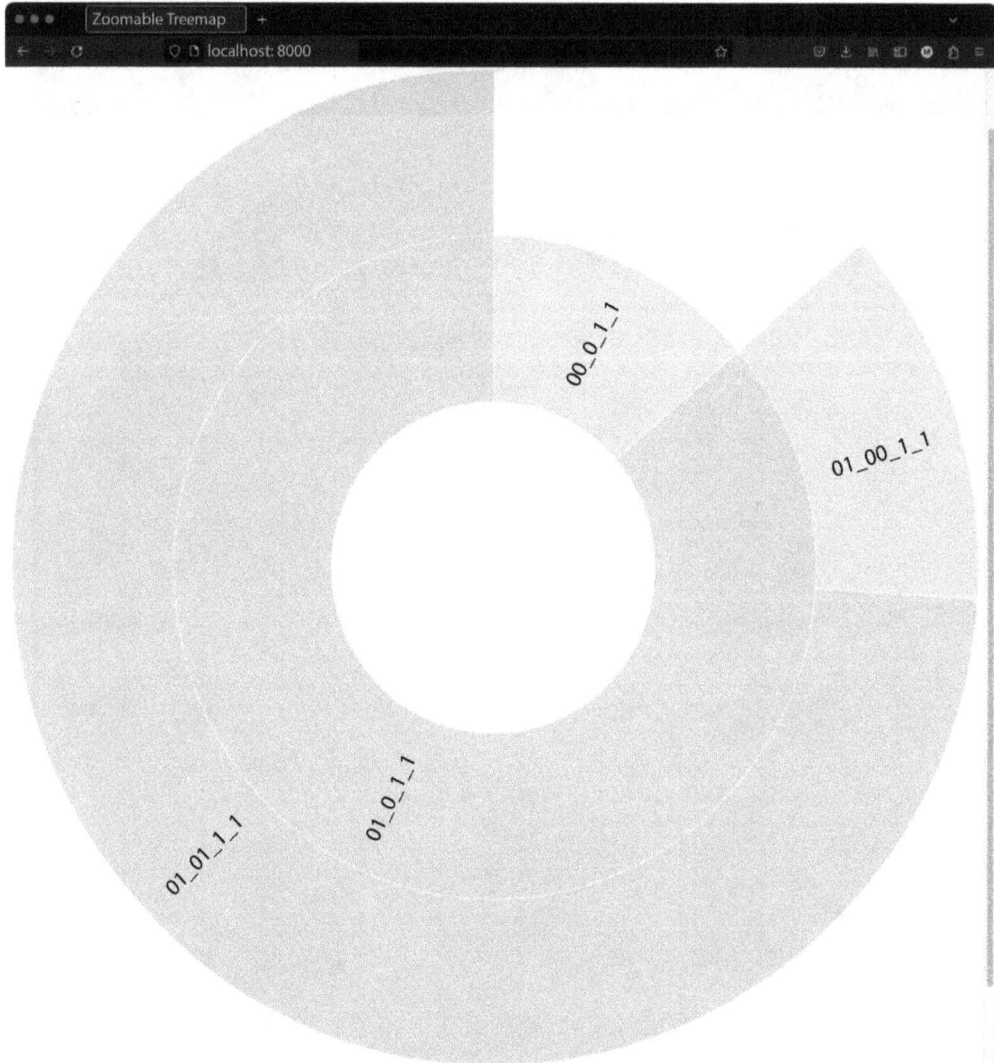

Figure 6.8 – Zooming in the 0_0_1_1 subtree of the Sunburst

Once again, the zooming capability is handy, and we want to have it in our visualizations. The next section discusses an interesting kind of plot, which is called icicle, and looks like it is suitable for visualizing iSAX indexes.

Using icicle plots

In this section, we are going to discuss a different kind of plot, which is called an **icicle plot**. An icicle plot is a method for presenting hierarchical clustering and is able to visualize hierarchical data using rectangular sectors that go from the root node to the leaves. In our case, we are going to use a **zoomable icicle plot**.

First, please go to the `ZoomableIcicle` directory and replace the JSON file in `files` with `100k. json`. This time, instead of changing the JavaScript code, we are going to change the field name of the JSON file from `size` to `value`. In general, *it is better to change your input data than the code*.

Figure 6.9 shows a part of the generated icicle visualization. The rectangle on the left side represents the root node, which contains **99,985** subsequences – this is the total number of subsequences stored in the iSAX index.

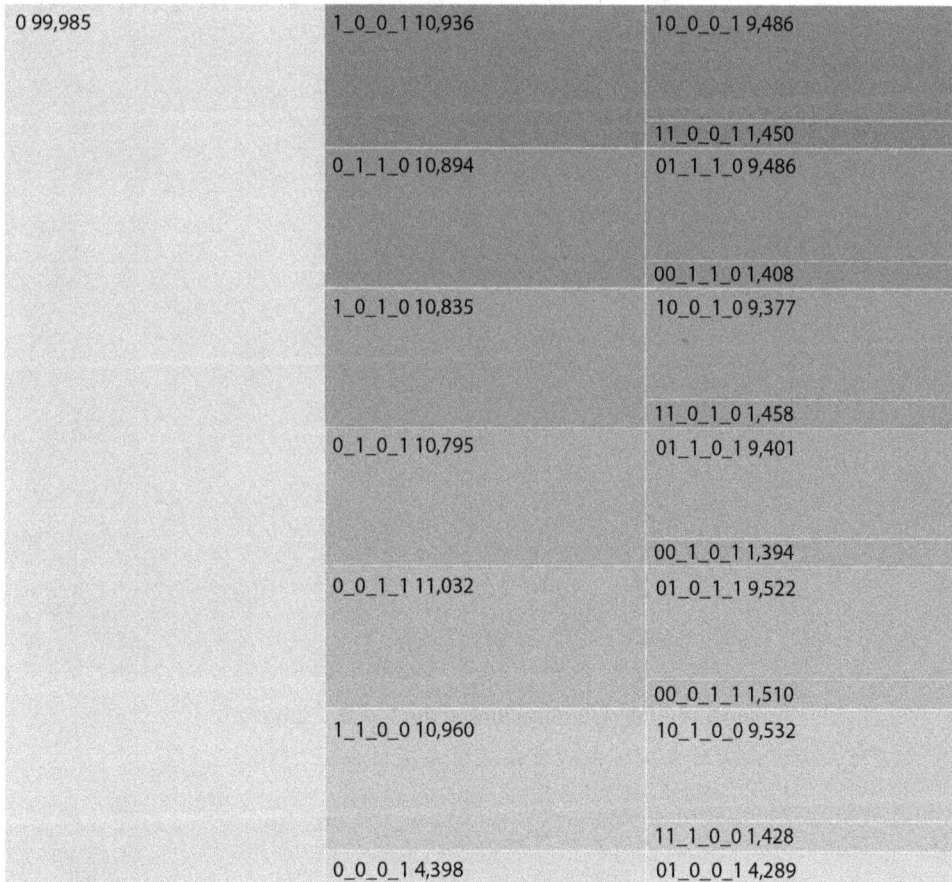

0 99,985	1_0_0_1 10,936	10_0_0_1 9,486
		11_0_0_1 1,450
	0_1_1_0 10,894	01_1_1_0 9,486
		00_1_1_0 1,408
	1_0_1_0 10,835	10_0_1_0 9,377
		11_0_1_0 1,458
	0_1_0_1 10,795	01_1_0_1 9,401
		00_1_0_1 1,394
	0_0_1_1 11,032	01_0_1_1 9,522
		00_0_1_1 1,510
	1_1_0_0 10,960	10_1_0_0 9,532
		11_1_0_0 1,428
	0_0_0_1 4,398	01_0_0_1 4,289

Figure 6.9 – Visualizing iSAX using an icicle

Apart from the SAX representation of a node, each rectangle displays the number of subsequences stored under it. So, the **1_0_0_1** subtree has **10,936** subsequences – this is another handy feature.

Going further, if we zoom on the **1_0_0_0** subtree, we are going to get the output displayed as shown in *Figure 6.10*:

Figure 6.10 – Taking a closer look at the 1_0_0_0 subtree

Similarly, if we zoom in on the **1_1_1_0** subtree, we are going to get the visualization presented in *Figure 6.11*:

1_1_1_0 4,342   10_1_1_0 4,246   10_10_1_0 4,133

10_11_1_0 113

11_1_1_0 96

Figure 6.11 – Taking a closer look at the 1_1_1_0 subtree

Let us discuss *Figure 6.11* a little more. What does it tell us? It tells us that the **1_1_1_0** child of the root node stores **4,342** subsequences. **4,246** of these subsequences are under the **10_1_1_0** subtree and the rest of the subsequences are under the **11_1_1_0** subtree.

If we zoom in on the **10_1_1_0** node, we are going to get *Figure 6.12*, which shows that the **10_10_10_0** subtree has **4,031** nodes.

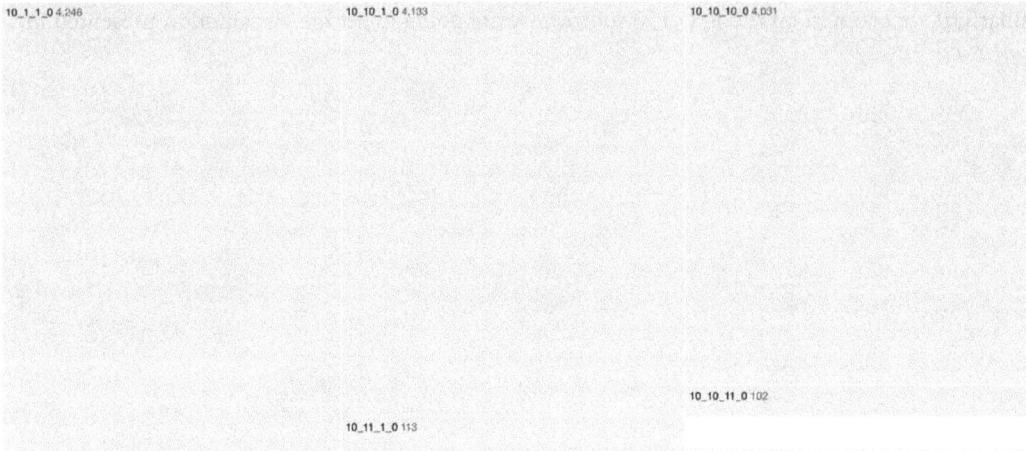

Figure 6.12 – Taking a closer look at the 10_1_1_0 subtree

As we are using a smaller threshold value, we know that the **10_10_10_0** node is an inner node that can be further expanded.

This way, we can explore the iSAX index and find the information we want.

The icicle plot looks appropriate for visualizing iSAX indexes. However, we might find a better type of visualization if we experiment more.

The next section presents a Collapsible Tree visualization of an iSAX index.

Visualizing iSAX as a Collapsible Tree

Although the zoomable icicle looks very promising, some people might want a visualization that looks like a tree but still has some of the versatility of the zoomable icicle. For those people, we are going to try the **Collapsible Tree**.

First, we go to the `CollapsibleTree` directory and then we run the Python web server. Then, we go to `http://localhost:8000/`. *Figure 6.13* shows the output of the Collapsible Tree visualization:

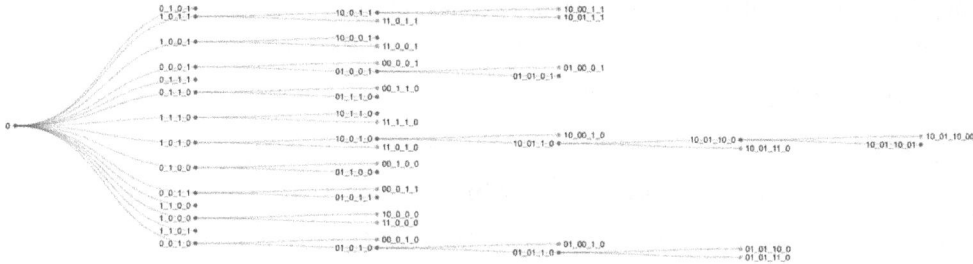

Figure 6.13 – Visualizing iSAX as a Collapsible Tree

The main advantage of the Collapsible Tree is that we can expand or collapse nodes at will, which means that we can easily concentrate on the nodes that interest us the most instead of getting lost in the details of the iSAX index.

However, the Collapsible Tree does not display the number of subsequences stored under each subtree of the index.

In this last section, we saw the operation of the Collapsible Tree and understood its versatility, as well as its limitations.

Summary

Visualization offers a great way to understand your data. Similarly, visualization is a great way to understand the structure of an iSAX index, especially a big one. In this chapter, we saw various ways to visualize an iSAX index with the help of the D3.js JavaScript library and got a better look at the distribution of the subsequences and the height of iSAX indexes.

However, it would be great to try your own visualizations using the D3.js JavaScript library, R, or other appropriate Python packages, which can also create impressive visualizations.

Lastly, do not underestimate the power of a good visualization as it can reveal lots of information in an easy-to-discover way. Just keep in mind that visualization is an art that is hard to master.

The next chapter is about using iSAX indexes for the approximate calculation of the Matrix Profile and the MPdist distance.

Useful links

- *JavaScript*: https://en.wikipedia.org/wiki/JavaScript
- The Mozilla Developer Network: https://developer.mozilla.org/en/JavaScript
- The official page for the D3.js JavaScript library: https://d3js.org/

- You can learn more information about icicle plots by reading the *Icicle Plots: Better Displays for Hierarchical Clustering* paper, which was written by J. B. Kruskal and J. M. Landwehr
- The R Project: `https://www.r-project.org/`
- The Seaborn Python package: `https://seaborn.pydata.org/`
- The Julia programming language: `https://julialang.org/`
- The `plotly` Python library: `https://plotly.com/python/`
- A very good book about the art of data visualization is *The Visual Display of Quantitative Information*, by Edward R. Tufte
- *D3 Gallery*: `https://observablehq.com/@d3/gallery`
- *Tree, Tidy*: `https://observablehq.com/@d3/tree`
- *Zoomable Treemap*: `https://observablehq.com/@d3/zoomable-treemap`
- *Zoomable Sunburst*: `https://observablehq.com/@d3/zoomable-sunburst`
- *Zoomable Icicle*: `https://observablehq.com/@d3/zoomable-icicle`
- *Tree, Radial Tidy*: `https://observablehq.com/@d3/radial-tree`
- *Collapsible Tree*: `https://observablehq.com/@d3/collapsible-tree`

Exercises

Try to do the following exercises:

- Create a time series with 50,000 elements and plot its iSAX index using 6, 8, and 10 segments. In all cases, use a threshold value of `500`.
- Create a time series with 150,000 elements and plot its iSAX index using 4, 6, and 8 segments.
- Create a time series with 250,000 elements and plot its iSAX index for 4, 6, and 10 segments. In all cases, use a threshold value of `5000`.
- Make a version of `exportJSON.py` that replaces the `size` field with a field named `value`.
- If you are familiar with JavaScript, change the colors of the zoomable icicle plot.
- If you are familiar with JavaScript, make the zoomable icicle plot go from *top to bottom* instead of *left to right*.
- If you are familiar with JavaScript, make the Collapsible Tree visualization go from *top to bottom* instead of *left to right*. Is that better than before?
- Experiment with the Zoomable Circle Packing visualization, which can be found at `https://observablehq.com/@d3/zoomable-circle-packing`. What do you think of it?

7

Using iSAX to Approximate MPdist

So far in this book, we have seen the use of iSAX for searching subsequences and joining iSAX indexes based on SAX representations but no other applications of it.

In this chapter, we are going to use iSAX indexes to approximately calculate the **Matrix Profile** vectors as well as the **MPdist** distance between two time series – we are still going to use iSAX for searching and joining, but the end results are going to be more sophisticated. The idea that governs this chapter is the perception that *terminal nodes in an iSAX index group have similar subsequences* from a SAX representation perspective – this is what we are trying to take advantage of for our approximate computations.

In this chapter, we are going to cover the following main topics:

- Understanding the Matrix Profile
- Computing the Matrix Profile using iSAX
- Understanding MPdist
- Calculating MPdist using iSAX
- Implementing the MPdist calculation in Python
- Using the Python code

Technical requirements

The GitHub repository for the book is at https://github.com/PacktPublishing/Time-Series-Indexing. The code for each chapter is in its own directory. Therefore, the code for *Chapter 7* can be found in the ch07 folder of the GitHub repository.

Understanding the Matrix Profile

Time series are everywhere, and there are many tasks that we might need to perform on large time series including similarity search, outlier detection, classification, and clustering. Dealing directly with a large time series is very time-consuming and is going to slow down the process. Most of the aforementioned tasks are based on the computation of the nearest neighbor of subsequences using a given sliding window size. This is where the **Matrix Profile** comes into play because it helps you perform the previous tasks once you have computed them.

We already saw the Matrix Profile in *Chapter 1*, but in this section, we are going to discuss it in more detail in order to understand better the reason that it is so slow to compute.

Various research papers exist that present and extend the Matrix Profile, including the following:

- *Matrix Profile I: All Pairs Similarity Joins for Time Series: A Unifying View That Includes Motifs, Discords and Shapelets*, written by Chin-Chia Michael Yeh, Yan Zhu, Liudmila Ulanova, Nurjahan Begum, Yifei Ding, Hoang Anh Dau, Diego Furtado Silva, Abdullah Mueen, and Eamonn J. Keogh (`https://ieeexplore.ieee.org/document/7837992`)

- *Matrix Profile II: Exploiting a Novel Algorithm and GPUs to Break the One Hundred Million Barrier for Time Series Motifs and Joins*, written by Yan Zhu, Zachary Zimmerman, Nader Shakibay Senobari, Chin-Chia Michael Yeh, Gareth Funning, Abdullah Mueen, Philip Brisk, and Eamonn Keogh (`https://ieeexplore.ieee.org/abstract/document/7837898`)

- *Matrix profile goes MAD: variable-length motif and discord discovery in data series*, written by Michele Linardi, Yan Zhu, Themis Palpanas, and Eamonn J. Keogh (`https://doi.org/10.1007/s10618-020-00685-w`)

> **About normalization**
>
> As it happens with the SAX representation, all Euclidean distances that are going to be computed in this chapter use normalized subsequences.

The next subsection shows what the Matrix Profile computation returns.

What does the Matrix Profile compute?

In this subsection, we are going to explain what the Matrix Profile calculates. Imagine having a time series and a sliding window size that is smaller than the time series length. The Matrix Profile computes *two vectors*.

The first vector contains the *Euclidean distances of the nearest neighbor* of each subsequence. The value at index 0 is the Euclidean distance of the nearest neighbor of the subsequence that begins at index 0, and so on.

In the second vector, the value at each place of the vector is the index of the subsequence that is the nearest neighbor and corresponds to the Euclidean distance stored in the previous vector. So, if the value at index 0 is 123, this means that the nearest neighbor of the subsequence that begins at index 0 in the original time series is the subsequence that begins at index 123 in the original time series. The first vector is going to contain that Euclidean distance value.

It is very important to understand that when computing the Matrix Profile for a time series using a self-join – that is, by looking for the nearest neighbor at the subsequences of the same time series – we need to *exclude the subsequences that are close* to the subsequence that we are examining. This is required because subsequences that share many elements in the same order tend to have smaller Euclidean distances by default. However, when dealing with a subsequence that is from another time series, we do not need to exclude any subsequences from the calculations.

A naïve implementation of the computation of the Matrix Profile vectors is to get the first subsequence, compare it to all other subsequences (excluding the subsequences that are close), find its nearest neighbor, and put the Euclidean distance and the index of the nearest neighbor at index 0 of the two vectors. Then, do the same for all other subsequences. Although this works for smaller time series, it is not very efficient as its algorithmic complexity is $O(n^2)$. This means that for a time series with 10,000 subsequences, we need to perform 10,000 times 10,000 computations (100,000,000). We are going to implement that algorithm to understand how slow it can be in real life.

The authors of the original Matrix Profile paper created a clever technique that involves **Fast Fourier** transforms that compute the Matrix Profile vectors with a viable complexity – the name of the algorithm is **Mueen's Algorithm for Similarity Search** (**MASS**). If you want to learn more about the details of the MASS algorithm and the ideas behind the Matrix Profile, you should read the *Matrix Profile I: All Pairs Similarity Joins for Time Series: A Unifying View That Includes Motifs, Discords and Shapelets* paper (https://ieeexplore.ieee.org/document/7837992).

The next section presents an implementation of the naïve algorithm for computing the Matrix Profile vectors. The naivete of the algorithm lies in its complexity, not its accuracy.

Manually computing the exact Matrix Profile

In this subsection, we are going to manually compute the exact Matrix Profile to show how slow the process can be, especially when working with large time series. We are using the word *exact* to differentiate this from the approximate Matrix Profile computation that we are going to implement in the *Computing Matrix Profile using iSAX* section of this chapter.

The last Python statements in the main() function of mp.py are the following:

```
dist, index = mp(ta, windowSize)
print(dist)
print(index)
```

The first statement runs the mp() function, which returns two values, both of them being lists (vectors), which are the two Matrix Profile vectors.

The implementation of the mp() function is where we compute the two vectors and is presented in two parts. The first part comes with the following code:

```
def mp(ts, window):
    l = len(ts) - window + 1
    dist = [None] * l
    index = [None] * l

    for i1 in range(l):
        t1 = ts[i1:i1+window]
        min = None
        minIndex = 0

        exclusionMin = i1 - window // 4
        if exclusionMin < 0:
            exclusionMin = 0

        exclusionMax = i1 + window // 4
        if exclusionMax > l-1:
            exclusionMax = l-1
```

In the previous code, we iterate over all the subsequences of the given time series. For each such subsequence, we define the indexes of the exclusion zone as specified in the *Matrix Profile I: All Pairs Similarity Joins for Time Series: A Unifying View That Includes Motifs, Discords and Shapelets* paper.

For a sliding window size of 16, the exclusion zone is 4 elements (16 // 4) on the left and 4 elements (16 // 4) on the right side of the subsequence.

The second part of mp() is as follows:

```
    for i2 in range(l):
        # Exclusion zone
        if i2 >= exclusionMin and i2 <= exclusionMax:
            continue

        t2 = ts[i2:i2+window]
        temp = round(euclidean(t1, t2), 3)
        if min == None:
            min = temp
            minIndex = i2
        elif min > temp:
            min = temp
```

```
                minIndex = i2

        dist[i1] = min
        index[i1] = minIndex

    return dist, index
```

In this part of the section, we compare each subsequence from the first part of the code with all the subsequences of the time series while taking into account the exclusion zone.

The bad thing about mp() is that it contains two for loops, which makes its computational complexity $O(n^2)$.

The output of mp.py when working with the ts.gz time series from *Chapter 6* (which is found in the ch06 directory of the GitHub repository of the book) is similar to the following for a sliding window of 16 – we are going to use the output to test the correctness of our implementation by comparing it to the original Matrix Profile algorithm and its output:

```
$ ./mp.py ../ch06/ts.gz -w 16
TS: ../ch06/ts.gz Sliding Window size: 16
[3.294, 3.111, 3.321, 3.535, 3.285, 3.373, 3.332, 3.693, 4.066, 4.065,
3.898, 3.484, 3.372, 3.1, 3.047, 3.299, 3.056, 3.361, 3.766, 3.759,
3.871, 3.884, 3.619, 3.035, 2.358, 3.012, 3.052, 3.136, 3.161, 3.219,
3.309, 3.526, 3.386, 3.973, 4.207, 4.101, 4.249, 4.498, 4.492, 4.255,
4.241, 3.285, 3.517, 3.494, 3.257, 3.316, 3.526, 4.183, 4.011, 3.294,
3.111, 3.321, 3.535, 3.1, 3.047, 3.332, 3.035, 2.358, 3.012, 3.052,
3.136, 3.161, 3.219, 3.201, 3.187, 3.017, 2.676, 2.763, 2.959, 3.952,
3.865, 3.678, 3.687, 3.201, 3.187, 3.017, 2.676, 2.763, 2.959, 3.316,
3.526, 3.899, 3.651, 3.664, 3.885]
[49, 50, 51, 52, 53, 54, 55, 56, 57, 46, 65, 27, 28, 53, 54, 74, 75,
76, 77, 59, 60, 61, 55, 56, 57, 58, 59, 60, 61, 62, 14, 15, 16, 66,
71, 68, 69, 56, 20, 63, 26, 75, 66, 67, 78, 79, 80, 81, 82, 0, 1, 2,
3, 13, 14, 6, 23, 24, 25, 26, 27, 28, 29, 73, 74, 75, 76, 77, 78, 79,
80, 61, 62, 63, 64, 65, 66, 67, 68, 45, 46, 62, 63, 64, 65]
--- 0.36465 seconds ---
```

Naively thinking, the subsequences with the smallest and largest Euclidean distances can be considered outliers as they differ from all other subsequences – this is an example of the use of the Matrix Profile for anomaly detection.

Using a sliding window size of 32, mp.py produces the following kind of output:

```
$ ./mp.py ../ch06/ts.gz -w 32
TS: ../ch06/ts.gz Sliding Window size: 32
[4.976, 5.131, 5.38, 5.485, 5.636, 5.75, 5.87, 6.076, 6.502, 6.705,
6.552, 6.145, 6.279, 6.599, 6.766, 6.667, 6.577, 6.429, 6.358, 6.358,
5.978, 5.804, 5.588, 5.092, 4.976, 5.01, 5.35, 5.456, 6.036, 6.082,
6.258, 6.513, 6.556, 6.553, 6.672, 6.745, 6.767, 6.777, 7.018, 7.12,
6.564, 6.203, 6.291, 6.118, 6.048, 5.869, 6.142, 6.431, 6.646, 4.976,
```

```
5.131, 5.38, 5.485, 5.636, 5.75, 5.588, 5.092, 4.976, 5.01, 5.35,
5.456, 6.036, 6.082, 6.258, 6.513, 6.556, 6.598, 6.518, 6.473]
[49, 50, 51, 52, 53, 54, 55, 56, 24, 58, 24, 25, 26, 27, 63, 64, 65,
55, 56, 52, 53, 54, 55, 56, 57, 58, 59, 60, 61, 62, 63, 64, 65, 67,
68, 62, 63, 0, 65, 67, 0, 1, 2, 3, 4, 5, 6, 7, 8, 0, 1, 2, 3, 4, 5,
22, 23, 24, 25, 26, 27, 28, 29, 30, 31, 32, 17, 57, 59]
--- 0.22118 seconds ---
```

Lastly, using a sliding window size of 64, the produced output is the following:

```
$ ./mp.py ../ch06/ts.gz -w 64
TS: ../ch06/ts.gz Sliding Window size: 64
[10.529, 10.406, 10.475, 10.377, 10.702, 10.869, 10.793, 10.827,
10.743, 11.14, 10.865, 10.819, 10.876, 10.808, 10.802, 10.73, 10.713,
10.67, 11.288, 11.296, 11.113, 11.202, 11.196, 11.121, 11.033, 11.145,
11.228, 11.125, 11.108, 10.865, 10.819, 10.671, 10.702, 10.529,
10.406, 10.475, 10.377]
[33, 34, 35, 36, 32, 33, 34, 35, 36, 28, 29, 30, 31, 32, 33, 34, 35,
36, 0, 1, 2, 3, 4, 5, 6, 7, 8, 8, 0, 10, 11, 3, 4, 0, 1, 2, 3]
--- 0.03179 seconds ---
```

The reason for having a smaller output here is that the bigger the sliding window size, the fewer the number of subsequences that are created from a time series.

Now, let us experiment with a time series with 25,000 elements that was created as follows:

```
$ ../ch01/synthetic_data.py 25000 -5 5 > 25k
$ gzip 25k
```

The results for 25k.gz with the same sliding window sizes as before are as follows (only the times are shown – the rest of the output is omitted for brevity):

```
$ ./mp.py -w 16 25k.gz
--- 43707.95353 seconds ---
$ ./mp.py -w 32 25k.gz
--- 44162.44419 seconds ---
$ ./mp.py -w 64 25k.gz
--- 45113.62417 seconds ---
```

At this point, we should be aware of the fact that computing the Matrix Profile vectors can be really slow as it took mp.py 45,113 seconds to compute the Matrix Profile in the last run.

Can you think of the reason that even a small increase in the sliding window size also increases the overall times? The answer is that the bigger the sliding window size, the bigger the subsequence length, and therefore, the more time it takes to compute the Euclidean distance between two subsequences. Here is the time it takes to compute the Matrix Profile vectors for a sliding window size of 2048:

```
$ ./mp.py -w 2048 25k.gz
--- 46271.63763 seconds ---
```

Have in mind that *the MASS algorithm does not have such an issue* as it computes the Euclidean distances in its own clever way. As a result, its performance depends on the time series length only.

Now, let us present a Python script that computes the exact Matrix Profile using the MASS algorithm with the help of the `stumpy` Python package. We are using the `realMP.py` script for computing the Matrix Profile vectors, which has the following implementation:

```python
#!/usr/bin/env python

import pandas as pd
import argparse
import time
import stumpy
import numpy as np

def main():
    parser = argparse.ArgumentParser()
    parser.add_argument("-w", "--window", dest = "window",
        default = "16", help="Sliding Window", type=int)
    parser.add_argument("TS")
    args = parser.parse_args()

    windowSize = args.window
    inputTS = args.TS
    print("TS:", inputTS, "Sliding Window size:",
        windowSize)

    start_time = time.time()
    ts = pd.read_csv(inputTS, names=['values'],
        compression='gzip')

    ts_numpy = ts.to_numpy()
    ta = ts_numpy.reshape(len(ts_numpy))
    realMP = stumpy.stump(ta, windowSize)

    realDistances = realMP[:,0]
```

```
            realIndexes = realMP[:,1]
            print("--- %.5f seconds ---" % (time.time() -
                start_time))

            print(realDistances)
            print(realIndexes)

if __name__ == '__main__':
    main()
```

The return value of `stumpy.stump()` is a multi-dimensional array. The first column (`[:,0]`) is the vector of distances, and the second column (`[:,1]`) is the vector of indexes. In the previous code, we print both these vectors, which is not very handy when dealing with large time series – comment out these two `print()` statements if you want.

In order to verify the correctness of `mp.py`, we present the output of `realMP.py` for the `ts.gz` time series and a sliding window size of 64:

```
$ ./realMP.py ../ch06/ts.gz -w 64
TS: ../ch06/ts.gz Sliding Window size: 64
--- 11.31371 seconds ---
[10.5292 10.40594 10.47460 10.3770 10.7024 10.8689 10.7928
 10.8274 10.74260 11.140 10.864 10.818 10.8757 10.8078
 10.8017 10.7296 10.7129 10.6704
 11.2882 11.2963 11.1125 11.2019 11.19556 11.1206
 11.0330 11.14458 11.22779 11.12475
 11.10825 10.864619 10.8186 10.6714
 10.7024 10.52926 10.40594 10.4746 10.3770]
[33 34 35 36 32 33 34 35 36 28 29 30 31 32 33 34 35 36 0 1 2 3 4 5 6 7
 8 8 0 10 11 3 4 0 1 2 3]
```

Now that we are sure about the correctness of `mp.py`, let us experiment with the `25k.gz` time series to see how much time it takes to compute the exact Matrix Profile vectors.

The time it takes `realMP.py` and the `stumpy.stump()` function to compute the Matrix Profile vectors *on a single CPU core* for the `25k.gz` time series is the following:

```
$ taskset --cpu-list 0 ./realMP.py 25k.gz -w 1024
TS: 25k.gz Sliding Window size: 1024
--- 11.19547 seconds ---
[42.41325061659 42.4212959655 42.45021115618 ...
 42.64248908665 42.64380072599 42.6591584368]
[10218 10219 10220 ... 7240 7241 20243]
```

The time it takes `realMP.py` to compute the Matrix Profile vectors for the `25k.gz` time series on an Intel i7 with 8 CPU cores is the following:

```
$ ./realMP.py 25k.gz -w 1024
TS: 25k.gz Sliding Window size: 1024
--- 9.68259 seconds ---
[42.41325061659 42.4212959655 42.45021115618 ...
 42.64248908665 42.64380072599 42.6591584368]
[10218 10219 10220 ... 7240 7241 20243]
```

Moreover, the time it takes `realMP.py` and the `stumpy.stump()` function to compute the Matrix Profile vectors *on a single CPU core* for the `ch06/100k.gz` time series and sliding window size of `1024` is the following:

```
$ taskset --cpu-list 0 ./realMP.py ../ch06/100k.gz -w 1024
TS: ../ch06/100k.gz Sliding Window size: 1024
--- 44.45451 seconds ---
[42.0661718111 42.044733861 42.050637591 ...
 42.252694931 42.225343182 42.2147590858]
[51861 51862 51863 ... 13502 13503 13504]
```

Lastly, let us try `realMP.py` on a *single CPU core* on the `500k.gz` time series from *Chapter 4*:

```
$ taskset --cpu-list 0 ./realMP.py ../ch04/500k.gz -w 1024
TS: ../ch04/500k.gz Sliding Window size: 1024
--- 1229.49608 seconds ---
[41.691930926 41.689248432 41.642429848 ...
 41.712625718 41.6520521157 41.636642904]
[446724 446725 446726 ... 260568 260569 260570]
```

The conclusion from the previous output is that computing the Matrix Profile gets slower as the length of the time series gets bigger, which is the main reason for thinking about an approximate computation of it. What we lose in accuracy, we gain in time. We cannot have everything!

The next section explains the technique that we are going to use to approximately compute the Matrix Profile vectors with the help of iSAX.

Computing the Matrix Profile using iSAX

First of all, let us make something clear: we are going to present *an approximate method*. If you want to calculate the exact Matrix Profile, then you should use an implementation that uses the original algorithm.

The idea behind the used technique is the following: *it is more likely that the nearest neighbor of a subsequence is going to be found in the subsequences stored in the same terminal node as the subsequence under examination*. Therefore, we do not need to check all the subsequences of the time series, just a small subset of them.

The next subsection discusses and resolves an issue that might come up in our calculations, which is what are we going to do if we cannot find a proper match for a subsequence in a terminal node.

What happens if there is not a valid match?

In this subsection, we are going to clarify the problematic cases of the process. There exist two conditions that might end up in an undesired situation:

- A terminal node contains a single subsequence only
- For a given subsequence, all the remaining subsequences of the terminal node are in the exclusion zone

In both cases, we are not going to be able to find the approximate nearest neighbor of a subsequence. Can we resolve these issues?

There exist multiple answers to that question, including doing nothing or choosing a different subsequence and using that to compute the Euclidean distance of the nearest neighbor. We are going with the latter solution, but instead of randomly choosing a subsequence, we are going for the subsequence that is next to the left side of the exclusion zone. If there is no space on the left side of the exclusion zone, we are going to choose the subsequence that is next to the right side of the exclusion zone. As these two conditions cannot happen at the same time, we are good!

The next subsection discusses how to compute the error of the approximate Matrix Profile vector of distances compared to the real Matrix Profile vector of distances.

Calculating the error

As explained earlier, we are computing an approximate Matrix Profile vector. In such cases, we need a way to compute how far we are from the real values. There exist various ways to compute an error value between two quantities. As a Matrix Profile is a list of values, we need to find a way to compute an error value that supports a list of values, not single values only.

The most common way is to find the Euclidean distance between the approximate vector and the exact vector. However, this does not always tell the whole truth. A good alternative would be to use the **Root Mean Square Error (RMSE)**.

The formula for the RMSE is a little complex at first. It is presented in *Figure 7.1*:

$$RMSE = \sqrt{\frac{\sum_i^n (approximate_i - real_i)^2}{n}}$$

Figure 7.1 – The RMSE formula

In practice, this means that we find the difference between the actual value and the approximate one and we square that. We do that for all the pairs and then add all these values – this is the purpose of the big Greek Sigma letter. After that, we divide by the number of the pairs. Lastly, we find the square root of that last value and we are done. If you are not good at mathematics, bear in mind that you do not need to remember that formula – we are going to implement it in Python in a while.

The desired property that the RMSE has is that it takes into account the number of elements that we compare. Put simply, the RMSE takes the *average*, whereas the Euclidean distance takes the *sum*. In our case, using the average error looks more appropriate.

As an example, the Euclidean distance between (0, 0, 0, 2, 2) and (2, 1, 0, 0, 0) is equal to 3.6055. On the other hand, the RMSE of these two vectors is equal to 1.61245.

With all that in mind, we are ready to present our approximate implementation.

Approximate Matrix Profile implementation

In this subsection, we present the Python script that approximately computes the Matrix Profile vectors.

The important code for apprMP.py can be found in approximateMP(), which is presented in four parts. The first part of the function is the following:

```
def approximateMP(ts_numpy):
    ISAX = isax.iSAX()

    length = len(ts_numpy)
    windowSize = variables.slidingWindowSize
    segments = variables.segments

    # Split sequence into subsequences
    for i in range(length - windowSize + 1):
        ts = ts_numpy[i:i+windowSize]
        ts_node = isax.TS(ts, segments)
```

```
        ts_node.index = i
        ISAX.insert(ts_node)

    vDist = [None] * (length - windowSize + 1)
    vIndex = [None] * (length - windowSize + 1)
    nSubsequences = length - windowSize + 1
```

The previous code splits the time series into subsequences and creates the iSAX index. It also initializes the vDist and vIndex variables, for keeping the list of distances and the list of indexes, respectively.

The second part of approximateMP() is the following:

```
for k in ISAX.ht:
        t = ISAX.ht[k]
        if t.terminalNode == False:
            continue

        # I is the index of the subsequence
        # in the terminal node
        for i in range(t.nTimeSeries()):
            # This is the REAL index of the subsequence
            # in the time series
            idx = t.children[i].index
            # This is the subsequence that we are examining
            currentTS = t.children[i].ts

            exclusionMin = idx-- windowSize // 4
            if exclusionMin < 0:
                exclusionMin = 0

            exclusionMax = idx + windowSize // 4
            if exclusionMax > nSubsequences-1:
                exclusionMax = nSubsequences-1

            min = None
            minIndex = 0
```

In the previous code, we take each node of the iSAX index and determine whether it is a terminal node or not – we are only interested in terminal nodes. If we are dealing with a terminal node, we process each subsequence stored there. First, we define the indexes of the exclusion zone making sure that the minimum value of the left side of the exclusion zone is 0 – this is the index of the first element of the time series – and the maximum value of the right side of the exclusion zone is not bigger than the length of the time series minus 1.

The third part of it is the following:

```
for sub in range(t.nTimeSeries()):
    # This is the REAL index of the subsequence
    # we are examining in the time series
    currentIdx = t.children[sub].index
    if currentIdx >= exclusionMin and currentIdx <=
exclusionMax:
        continue

    temp = round(tools.euclidean(currentTS,
        t.children[sub].ts), 3)
    if min == None:
        min = temp
        minIndex = currentIdx
    elif min > temp:
        min = temp
        minIndex = currentIdx
```

We compare each subsequence of the selected terminal node with the rest of the subsequences it contains because we expect that there is a high probability of the nearest neighbor being in the same node.

Then, we make sure that the index of the subsequence that is going to be compared with the initial subsequence is not in the exclusion zone. If we find such a subsequence, we compute the Euclidean distance and keep the relevant index value. From all these subsequences that are outside the exclusion zone and are located in the terminal node, we keep the minimum Euclidean distance and the related index.

We do that for all the subsequences in all the terminal nodes of the iSAX index.

The last part of the `approximateMP()` function is the following:

```
# Pick left limit first, then the right limit
if min == None:
    if exclusionMin-1 > 0:
        randomSub = ts_numpy[exclusionMin-
            1:exclusionMin+windowSize-1]
        vDist[idx] = round(tools.euclidean(
            currentTS, randomSub), 3)
        vIndex[idx] = exclusionMin - 1
    else:
        randomSub = ts_numpy[exclusionMax+
            1:exclusionMax+windowSize+1]
        vDist[idx] = round(tools.euclidean(
            currentTS, randomSub), 3)
        vIndex[idx] = exclusionMax + 1
```

```
            else:
                vDist[idx] = min
                vIndex[idx] = minIndex

    return vIndex, vDist
```

If, at this point, we do not have a valid Euclidean distance value (None), we compare the initial subsequence with the subsequence next to the left side of the exclusion zone, if it exists – this means if the left side of the exclusion zone is not 0. Otherwise, we compare it with the subsequence next to the right side of the exclusion zone. We put the relevant index and Euclidean distance into the vIndex and vDist variables, respectively. However, if we already have an index and Euclidean distance from earlier, we use these values.

The next subsection compares the accuracy of our approximate technique when using different iSAX parameters.

Comparing the accuracy of two different parameter sets

In this subsection, we are going to compute the approximate Matrix Profile vector of a single time series using two different sets of iSAX parameters and check the accuracy of the results using the RMSE.

To make things simpler, we have created a Python script that computes the two approximate Matrix Profile vectors of Euclidean distances, as well as the exact Matrix Profile vectors, and calculates the RMSE – the name of the script is rmse.py. We are not going to present the entire Python code of rmse.py, just the important Python statements, starting from the function that computes the RMSE:

```
def RMSE(realV, approximateV):
    diffrnce = np.subtract(realV, approximateV)
    sqre_err = np.square(diffrnce)
    rslt_meansqre_err = sqre_err.mean()
    error = math.sqrt(rslt_meansqre_err)

    return error
```

The previous code implements the computation of the RMSE value according to the formula presented in *Figure 7.1*.

The remaining relevant Python code is located in the main() function:

```
# Real Matrix Profile
TSreshape = ts_numpy.reshape(len(ts_numpy))
realMP = stumpy.stump(TSreshape, windowSize)
realDistances = realMP[:,0]


# Approximate Matrix Profile
```

```
    _, vDist = approximateMP(ts_numpy)

    rmseError = RMSE(realDistances, vDist)
    print("Error =", rmseError)
```

First, we compute the real Matrix Profile vector with `stumpy.stump()`, and then we compute the approximate Matrix Profile vector with the Euclidean distances using `approximateMP()`. After that, we call the `RMSE()` function and get the numeric result, which we print on the screen.

So, let us run `rmse.py` and see what we get:

```
$ ./rmse.py -w 32 -s 4 -t 500 -c 16 25k.gz
Max Cardinality: 16 Segments: 4 Sliding Window: 32 Threshold: 500
Default Promotion: False
Error = 10.70823863253679
```

Now, let us use `rmse.py` another time, but this time, with different iSAX parameters, as follows:

```
$ ./rmse.py -w 32 -s 4 -t 1500 -c 16 25k.gz
Max Cardinality: 16 Segments: 4 Sliding Window: 32 Threshold: 1500
Default Promotion: False
Error = 9.996114543048341
```

What do the previous results tell us? First, the results tell us that our approximate technique does not leave any subsequence without a Euclidean distance. If there was such a case, then `rmse.py` would have generated an error message like the following:

```
TypeError: unsupported operand type(s) for -: 'float' and 'NoneType'
```

As, in the initialization of `vDist`, all its elements are set equal to `None`, the previous error means that the value of at least one of the elements was not reset. Therefore, it is still equal to `None` and our code fails to subtract a floating point value, calculated by `stumpy.stump()`, from `None`.

Apart from that, the results tell us that bigger threshold values produce more accurate results, which makes perfect sense, as there are more subsequences in each terminal node. However, this makes the computation of the approximate Matrix Profile slower. As a general rule, the closer the number of subsequences at each terminal node is to the threshold value, the better the accuracy – we do not want terminal nodes with a small number of subsequences stored in them.

Now that we know about the Matrix Profile, let us discuss MPdist, how it is computed, and the role of the Matrix Profile in this computation.

Understanding MPdist

Now that we know about the Matrix Profile, we are ready to learn about MPdist and how the Matrix Profile is used in the calculation of MPdist. The paper that defines the MPdist distance is *Matrix Profile XII: MPdist: A Novel Time Series Distance Measure to Allow Data Mining in More Challenging Scenarios*, written by S. Gharghabi, S. Imani, A. Bagnall, A. Darvishzadeh, and E. Keogh (`https://ieeexplore.ieee.org/abstract/document/8594928`).

The intuition behind MPdist is that two time series can be considered similar if they *have similar patterns throughout their duration*. Such patterns are extracted in the form of subsequences using a sliding window. This is illustrated in *Figure 7.2*:

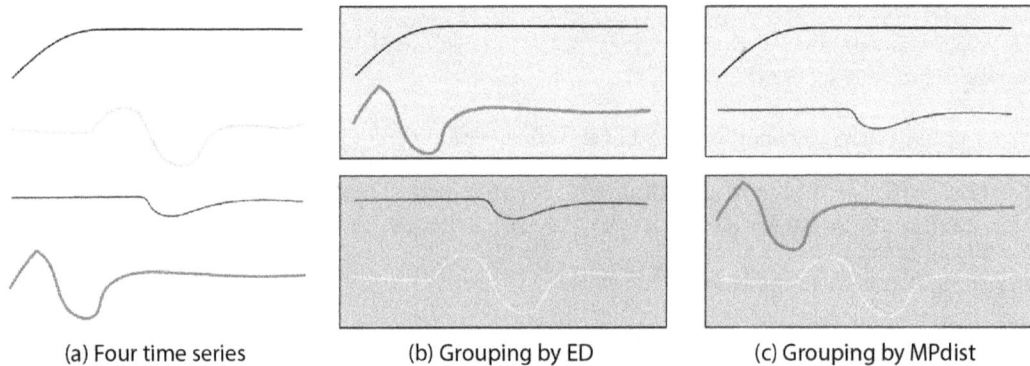

(a) Four time series   (b) Grouping by ED   (c) Grouping by MPdist

Figure 7.2 – Grouping time series

In *Figure 7.2*, we see that MPdist *(c)* understands the similarity between time series that follow the same pattern better, whereas Euclidean distance *(b)* compares time series based on time, and therefore groups the presented time series differently. In my opinion, the grouping that is based on MPdist is more accurate.

The advantages of MPdist (according to the people that created it) are that the MPdist distance measure tries to be more flexible than most available distance measures, including the Euclidean distance, and it takes into account similarities that may not take place at the same time. Additionally, MPdist can compare time series of different sizes –Euclidean distance cannot do that – and requires just a single parameter (the sliding window size) to operate.

The next subsection discusses the way MPdist is computed.

How to compute MPdist

In this subsection, we are going to discuss the way the real MPdist is computed in order to better understand the complexity of the process.

The computation of MPdist is based on the Matrix Profile. First, we are given two time series, A and B, and a sliding window size. Then, for each subsequence of the first time series, we find its nearest neighbor in the second time series, and we put the related Euclidean distance into a list of values. We do that for all the subsequences of the first time series. This is also called the **AB join**. Then, we do the same but for the second time series – this is called the **BA join**. So, in the end, we calculated the **ABBA join** and we have a list of Euclidean distances that we sort from the smallest to the biggest. From that list, we get the Euclidean distance found at the index value that is equal to *5% of the length of the list* – the authors of MPdist decided to use the Euclidean distance at that index as the MPdist value.

For both the AB join and BA join, the authors of MPdist use the MASS algorithm to compute the nearest neighbor of each subsequence, in order to avoid the inefficient algorithmic complexity of $O(n^2)$.

In the next subsection, we will create a Python script that manually computes the MPdist distance between two time series.

Manually computing MPdist

In this subsection, we are going to show how to manually compute the MPdist value between two time series. The idea behind the implementation is based on the code found in mp.py – however, fundamental differences exist as the Matrix Profile returns a vector of values instead of a single value.

The logic code of mpdist.py is implemented in two functions, named mpdist() and JOIN(). mpdist() is implemented as follows:

```
def mpdist(ts1, ts2, window):
    L_AB = JOIN(ts1, ts2, window)
    L_BA = JOIN(ts2, ts1, window)

    JABBA = L_AB + L_BA
    JABBA.sort()

    index = int(0.05 * (len(JABBA) + 2 * window)) + 1
    return JABBA[index]
```

The previous code uses the JOIN() function to compute AB Join and BA Join. Then, it concatenates the numeric results, which are all Euclidean distances, and sorts them. Based on the length of the concatenation, it computes index, which is used for selecting a value from the JABBA array.

JOIN() is implemented as follows:

```
def JOIN(ts1, ts2, window):
    LIST = []

    l1 = len(ts1) - window + 1
    l2 = len(ts2) - window + 1
```

```
    for i1 in range(11):
        t1 = ts1[i1:i1+window]
        min = round(euclidean(t1, ts2[0:window]), 4)
        for i2 in range(1, 12):
            t2 = ts2[i2:i2+window]
            temp = round(euclidean(t1, t2), 4)
            if min > temp:
                min = temp

        LIST.append(min)

    return LIST
```

This is where the join is implemented. For every subsequence in the ts1 time series, we find the nearest neighbor in the ts2 time series – there is no need for an exclusion zone in this case.

The bad thing about mpdist.py is that it contains two for loops, which makes its computational complexity $O(n^2)$ – this is no surprise, as MPdist is based on the Matrix Profile. Therefore, the previous technique is viable for small time series only. In general, **brute-force algorithms** usually do not work well for large amounts of data.

At this point, we are going to create two time series with 10,000 elements each:

```
$ ../ch01/synthetic_data.py 10000 -5 5 > 10k1
$ ../ch01/synthetic_data.py 10000 -5 5 > 10k2
$ gzip 10k1; gzip 10k2
```

The output of mpdist.py when working with 10k1.gz and 10k2.gz and a sliding window size of 128 is as follows:

```
$ ./mpdist.py 10k1.gz 10k2.gz -w 128
--- 12026.64167 seconds ---
MPdist: 12.5796
```

It took mpdist.py approximately 12,026 seconds to compute MPdist.

The output of mpdist.py when working with 10k1.gz and 10k2.gz and a sliding window size of 2048 is the following:

```
$ ./mpdist.py 10k1.gz 10k2.gz -w 2048
--- 9154.55179 seconds ---
MPdist: 60.7277
```

Why do you think the calculation of the 2048 sliding window ran faster than the same calculation for the sliding window size of 128? It most likely has to do with the fact that the 2048 sliding window needs fewer iterations (1,920 times 1,920, which is equal to 3,686,400) due to the larger sliding window size, which also compensates for the cost of computing Euclidean distances for larger subsequences in the 2048 sliding window case.

Let us now see how much time it takes the MASS algorithm to compute MPdist.

The time it takes the `stumpy.mpdist()` function to compute the previous MPdist distances on a single CPU core is the following – we are using the `mpdistance.py` script from *Chapter 1*:

```
$ taskset --cpu-list 0 ../ch01/mpdistance.py 10k1.gz 10k2.gz 128
TS1: 10k1.gz TS2: 10k2.gz Window Size: 128
--- 10.28342 seconds ---
MPdist: 12.5790
$ taskset --cpu-list 0 ../ch01/mpdistance.py 10k1.gz 10k2.gz 2048
TS1: 10k1.gz TS2: 10k2.gz Window Size: 2048
--- 10.03479 seconds ---
MPdist: 60.7277
```

So, it takes the `stumpy.mpdist()` function about 10 seconds.

The time it takes the `stumpy.mpdist()` function to compute the previous MPdist distances on four CPU cores is the following:

```
$ taskset --cpu-list 0,1,2,3 ../ch01/mpdistance.py 10k1.gz 10k2.gz 128
TS1: 10k1.gz TS2: 10k2.gz Window Size: 128
--- 9.42861 seconds ---
MPdist: 12.5790
$ taskset --cpu-list 0,1,2,3 ../ch01/mpdistance.py 10k1.gz 10k2.gz
2048
TS1: 10k1.gz TS2: 10k2.gz Window Size: 2048
--- 9.33578 seconds ---
MPdist: 60.7277
```

Why are the times almost the same when using a single CPU core? The answer is that with small time series, `stumpy.mpdist()` *does not have enough time* to use all CPU cores.

Lastly, the time it takes the `stumpy.mpdist()` function to compute the two MPdist distances on eight CPU cores is the following:

```
$ ../ch01/mpdistance.py 10k1.gz 10k2.gz 128
TS1: 10k1.gz TS2: 10k2.gz Window Size: 128
--- 9.54642 seconds ---
MPdist: 12.5790
$ ../ch01/mpdistance.py 10k1.gz 10k2.gz 2048
```

```
TS1: 10k1.gz TS2: 10k2.gz Window Size: 2048
--- 9.33648 seconds ---
MPdist: 60.7277
```

Why are the times the same when using four CPU cores? As before, for very small time series, the number of CPU cores used does not make any difference to the computation time as there is not enough time to use them.

We are now ready to use the existing knowledge to approximately compute MPdist with the help of iSAX.

Calculating MPdist using iSAX

In this section, we are going to discuss our views and ideas regarding using iSAX indexes to *approximately compute MPdist*.

We know that iSAX keeps together subsequences with the same SAX representation. As before, our feeling is that it is more likely to find the nearest neighbor of a subsequence from a given time series in the subsequences with the same SAX representation from another time series.

The next section is about putting our thoughts into practice.

Implementing the MPdist calculation in Python

In this section, we will discuss two ways to approximately compute MPdist with the help of iSAX.

The first way is much simpler than the second one and is slightly based on the approximate calculation of the Matrix Profile. We take each subsequence from the first time series, and we match it with a terminal node with the same SAX representation from the iSAX index of the second time series in order to get the approximate nearest neighbor – if a subsequence does not have a match *based on its SAX representation*, we ignore that subsequence. So, in this case, we do not join iSAX indexes, which makes the process much slower – our experiments are going to show how much slower this technique is.

For the second way, we just use the **similarity join** of two iSAX indexes, which we first saw in *Chapter 5*.

The next subsection shows the implementation of the first technique.

Using the approximate Matrix Profile way

Although we do not return any Matrix Profile vectors, this technique looks like computing the Matrix Profile because *we examine subsequences one by one* and not in groups, and return their Euclidean distance with the approximate nearest neighbor. In this technique, *there is no exclusion zone* in the computation because we are comparing subsequences from two different time series.

The important code within `apprMPdist.py` is the following – we assume that we have already generated the two iSAX indexes:

```
# We search iSAX2 for the NN of the
# subsequences from TS1
for idx in range(0, len(ts1)-windowSize+1):
    currentQuery = ts1[idx:idx+windowSize]
    t = NN(i2, currentQuery)
    if t != None:
        ED.append(t)


# We search iSAX1 for the NN of the
# subsequences from TS2
for idx in range(0, len(ts2)-windowSize+1):
    currentQuery = ts2[idx:idx+windowSize]
    t = NN(i1, currentQuery)
    if t != None:
        ED.append(t)


ED.sort()
idx = int(0.05 * ( len(ED) + 2 * windowSize)) + 1
print("Approximate MPdist:", round(ED[idx], 3))
```

For each subsequence of the first time series, search the iSAX index of the second time series for the approximate nearest neighbor using the NN() function. Then, do the same for the subsequences of the second time series and the iSAX index of the first time series.

What is interesting is the implementation of the NN() function, used in the previous code. We are going to present NN() in three parts. The first part is the following:

```
def NN(ISAX, q):
    ED = None
    segments = variables.segments
    threshold = variables.threshold

    # Create TS Node
    qTS = isax.TS(q, segments)

    segs = [1] * segments
    # If the relevant child of root is not there
    # we have a miss
    lower_cardinality = tools.lowerCardinality(segs, qTS)

    lower_cardinality_str = ""
```

```
for i in lower_cardinality:
    lower_cardinality_str=lower_cardinality_str+"_"+i

lower_cardinality_str = lower_cardinality_str[1:len(
    lower_cardinality_str)]
if ISAX.ht.get(lower_cardinality_str) == None:
    return None
```

In the previous code, we try to find an iSAX node with the same SAX representation as the subsequence we are examining – we begin with the children of the root node of the iSAX. If such a child of the root node cannot be found, then we have a miss and we ignore that particular subsequence. As the final list of Euclidean distances is large (this depends on the lengths of the time series), missing some subsequences has no real effect on the end result.

The second part of NN() is the following:

```
# Otherwise, we have a hit
n = ISAX.ht.get(lower_cardinality_str)
while n.terminalNode == False:
    left = n.left
    right = n.right

    leftSegs = left.word.split('_')
    # Promote
    tempCard = tools.promote(qTS, leftSegs)

    if tempCard == left.word:
        n = left
    elif tempCard == right.word:
        n = right
```

In the previous code, we try to locate the iSAX node with the desired SAX representation by traversing the iSAX index.

The last part of NN() is the following:

```
# Iterate over the subsequences of the terminal node
for i in range(0, threshold):
    child = n.children[i]
    if type(child) == isax.TS:
        distance = tools.euclidean(normalize(child.ts),
            normalize(qTS.ts))
        if ED == None:
            ED = distance
        if ED > distance:
```

```
                ED = distance
        else:
            break


    return ED
```

After locating the desired terminal node, we compare its subsequences with the given subsequence and return the minimum Euclidean distance found. The main program puts all these minimum Euclidean distances into a list.

Now, let us discuss the second technique, which joins two iSAX indexes.

Using the join of two iSAX indexes

The second way is much faster than the first one. In this way, we *join the two iSAX indexes* based on the technique from *Chapter 5*, and we get the list of Euclidean distances. From that list, we choose a value to be the approximate MPdist.

> **What happens if there is not a match among iSAX nodes?**
>
> In some rare cases that depend on the time series data and the iSAX parameters, some nodes from one iSAX might end up not having a match in the other iSAX, and vice versa. In our case, we *ignore those nodes*, which means that we end up having a smaller-than-expected list of Euclidean distances.

The important code within joinMPdist.py is the following – we assume that we have already generated the two iSAX indexes:

```
# Join the two iSAX indexes
Join(i1, i2)
variables.ED.sort()

print("variables.ED length:", len(variables.ED))

# Index
idx = int(0.05*(len(variables.ED) + 2*windowSize))+1
print("Approximate MPdist:", variables.ED[idx])
```

The previous code uses the Join() function from isax.iSAXjoin, which we implemented and saw in *Chapter 5*. We have already seen the join of two iSAX indexes. However, this is the first time that we actually use the results of that join for something.

We are now going to start using the existing implementations and see their performance.

Using the Python code

In this section, we are going to use the Python scripts that we have created.

Running `apprMPdist.py` using the two time series with 10,000 elements each that we created earlier in this chapter generates the following kind of output:

```
$ ./apprMPdist.py 10k1.gz 10k2.gz -s 3 -c 64 -t 500 -w 120
Max Cardinality: 64 Segments: 3 Sliding Window: 120 Threshold: 500
Default Promotion: False
MPdist: 351.27 seconds
Approximate MPdist: 12.603
```

Using a bigger sliding window size generates the following output:

```
$ ./apprMPdist.py 10k1.gz 10k2.gz -s 3 -c 64 -t 500 -w 300
Max Cardinality: 64 Segments: 3 Sliding Window: 300 Threshold: 500
Default Promotion: False
MPdist: 384.74 seconds
Approximate MPdist: 21.757
```

So, bigger sliding window sizes require more time. As before, this is because calculating Euclidean distances for bigger sliding window sizes is slower.

Executing `joinMPdist.py` produces the following output:

```
$ ./joinMPdist.py 10k1.gz 10k2.gz -s 3 -c 64 -t 500 -w 120
Max Cardinality: 64 Segments: 3 Sliding Window: 120 Threshold: 500
Default Promotion: False
MPdist: 37.70 seconds
variables.ED length: 17605
Approximate MPdist: 12.60282
```

As before, using a bigger sliding window produces the following output:

```
$ ./joinMPdist.py 10k1.gz 10k2.gz -s 3 -c 64 -t 500 -w 300
Max Cardinality: 64 Segments: 3 Sliding Window: 300 Threshold: 500
Default Promotion: False
MPdist: 31.24 seconds
variables.ED length: 13972
Approximate MPdist: 21.76263
```

It looks like `joinMPdist.py` is a lot faster than `apprMPdist.py`, which makes perfect sense as it is *using two iSAX indexes at the same time* to construct the list of Euclidean distances. Put simply, the running of `joinMPdist.py` requires fewer computations.

The next subsection compares the accuracy and the speed of the two methods when working with larger time series.

Comparing the accuracy and the speed of the methods

Both methods are far from perfect. However, in this subsection, we are going to compare their accuracy and speed in relation to the MPdist implementation found in the `stumpy` Python package.

We want to test our code on bigger time series, as this is where our technique might be faster than the exact MPdist function of `stumpy`. In this case, we are going to use two time series with around 500,000 elements each – we already created such time series in *Chapter 5*.

For `apprMPdist.py`, the results for sliding window sizes of `120`, `600`, and `1200` are as follows:

```
$ ./apprMPdist.py ../ch05/500k.gz ../ch05/506k.gz -s 6 -c 64 -t 500 -w
120
Max Cardinality: 32 Segments: 6 Sliding Window: 120 Threshold: 500
Default Promotion: False
MPdist: 19329.64 seconds
Approximate MPdist: 12.405
$ ./apprMPdist.py ../ch05/500k.gz ../ch05/506k.gz -s 6 -c 64 -t 500 -w
600
Max Cardinality: 64 Segments: 6 Sliding Window: 600 Threshold: 500
Default Promotion: False
MPdist: 21219.60 seconds
Approximate MPdist: 31.871
$ ./apprMPdist.py ../ch05/500k.gz ../ch05/506k.gz -s 6 -c 64 -t 500 -w
1200
Max Cardinality: 64 Segments: 6 Sliding Window: 1200 Threshold: 500
Default Promotion: False
MPdist: 23120.07 seconds
Approximate MPdist: 46.279
```

For the `joinMPdist.py` script, the output for sliding window sizes of `120`, `600`, and `1200` is the following:

```
$ ./joinMPdist.py ../ch05/500k.gz ../ch05/506k.gz -s 6 -c 64 -t 500 -w
120
Max Cardinality: 64 Segments: 6 Sliding Window: 120 Threshold: 500
Default Promotion: False
MPdist: 2595.92 seconds
variables.ED length: 910854
Approximate MPdist: 12.40684
$ ./joinMPdist.py ../ch05/500k.gz ../ch05/506k.gz -s 6 -c 64 -t 500 -w
600;
Max Cardinality: 64 Segments: 6 Sliding Window: 600 Threshold: 500
Default Promotion: False
```

```
MPdist: 2270.72 seconds
variables.ED length: 798022
Approximate MPdist: 31.88064
$ ./joinMPdist.py ../ch05/500k.gz ../ch05/506k.gz -s 6 -c 64 -t 500 -w
1200
Max Cardinality: 64 Segments: 6 Sliding Window: 1200 Threshold: 500
Default Promotion: False
MPdist: 2145.76 seconds
variables.ED length: 674777
Approximate MPdist: 46.29538
```

The results of joinMPdist.py are really promising when working with larger time series. Although it looks like the bigger the sliding window size, the faster the technique, this is not completely true because as the sliding window gets bigger, we have more nodes without a match, and therefore, the list of values gets smaller, which means that we compute fewer Euclidean distances as the sliding window increases. This is not always the case, as this depends on the time series data.

Lastly, the result from the stumpy Python package when running on a single CPU core is as follows:

```
$ taskset --cpu-list 0 ../ch01/mpdistance.py ../ch05/500k.gz ../
ch05/506k.gz 120
TS1: ../ch05/500k.gz TS2: ../ch05/506k.gz Window Size: 120
500000 506218
--- 4052.73237 seconds ---
MPdist: 11.4175
$ taskset --cpu-list 0 ../ch01/mpdistance.py ../ch05/500k.gz ../
ch05/506k.gz 600
TS1: ../ch05/500k.gz TS2: ../ch05/506k.gz Window Size: 600
500000 506218
--- 4042.52154 seconds ---
MPdist: 30.7796
$ taskset --cpu-list 0 ../ch01/mpdistance.py ../ch05/500k.gz ../
ch05/506k.gz 1200
TS1: ../ch05/500k.gz TS2: ../ch05/506k.gz Window Size: 1200
500000 506218
--- 4045.72392 seconds ---
MPdist: 45.1887
```

Figure 7.3 shows the accuracy of the approximate methods, which are named **Search** and **Join**, compared to the real MPdist value, which is named **Real**, for the three sliding window sizes used.

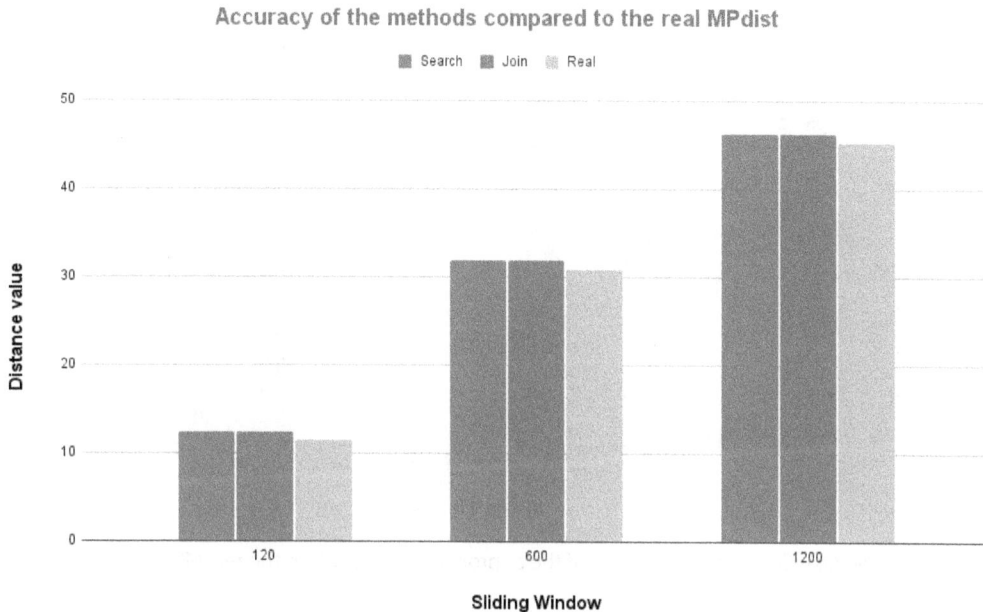

Figure 7.3 – Comparing the accuracy of the approximate methods to the real MPdist

What does the output of *Figure 7.3* tell us? First of all, the approximate methods performed pretty well because the approximate values are really close to the real MPdist values. So, at least for our example time series, the approximate techniques are very accurate.

Similarly, *Figure 7.4* compares the times of the approximate methods to the time of the `stumpy` computation when running on a single CPU core for the three sliding window sizes used – the presented times for the approximate methods *do not include the time it takes to create the two iSAX indexes*.

Times to run for each method

Search   Join   Real

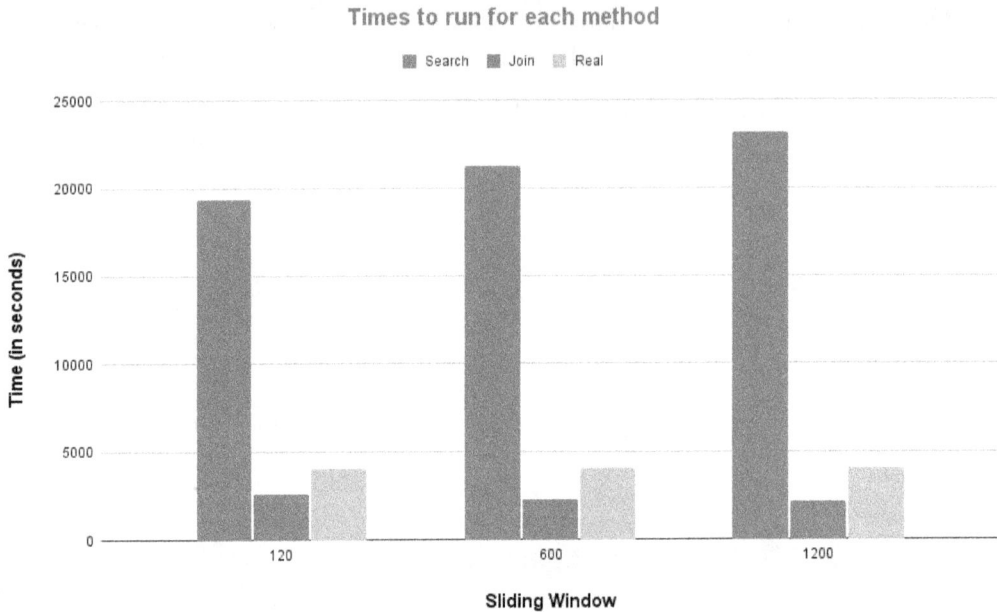

Figure 7.4 – Comparing the times of the approximate methods to the real MPdist

What does the output of *Figure 7.4* tell us? The first technique is really slow and should not be used – that is the purpose of experimentation: to find out what works well and what does not. On the other hand, the performance of the second approximate technique is very good. Additionally, *Figure 7.4* shows that the `stumpy` computation takes the same time regardless of the sliding window size – this is a good and desirable feature of the MASS algorithm.

Summary

Although the main purpose of iSAX is to help us search for subsequences by indexing them, there are other ways to use an iSAX index.

In this chapter, we presented a way to approximately compute the Matrix Profile vectors and two ways to approximately compute the MPdist distance between two time series. All these techniques use iSAX indexes.

We presented two ways to approximately compute MPdist. Out of the two methods, the one that joins two iSAX indexes is much more efficient than the other – so the use of an iSAX index by itself does not guarantee efficiency; we have to use an iSAX index the right way to get better results.

There is a small chapter left to finish this book, which is about the next steps you can follow if you are really into time series and databases.

Useful links

- The `stumpy` Python package: `https://pypi.org/project/stumpy/`
- The `numba` Python package: `https://pypi.org/project/numba/`
- The RMSE: `https://en.wikipedia.org/wiki/Root-mean-square_deviation`
- The UCR Matrix Profile page: `https://www.cs.ucr.edu/~eamonn/MatrixProfile.html`
- SAX home page: `https://www.cs.ucr.edu/~eamonn/SAX.htm`

Exercises

Try to do the following exercises:

- Try to use `mp.py` with a time series with 50,000 elements and see how much time it takes to complete for sliding window sizes of `16`, `2048`, and `4096`.
- Try to use `mp.py` with a time series with 65,000 elements and see how much time it takes to complete.
- Experiment with the exclusion zone limits of `mp.py` and see what you get.
- Use `realMP.py` and `stumpy.stump()` to compute the Matrix Profile vectors for a time series with 200,000 elements – create that time series if you do not have one.
- Use `realMP.py` and `stumpy.stump()` to compute the Matrix Profile vectors for a time series with 500,000 elements. Now, consider that a time series with 500,000 elements is on the small side!
- Try `realMP.py` on the `2M.gz` time series from *Chapter 4* using a single CPU code. As you can see, `realMP.py` starts getting really slow with larger time series. Now, consider that a time series with 2,000,000 elements is not big.
- We can make `mp.py` a little faster by storing the normalized versions of the subsequences and using the normalized versions when calculating the Euclidean distances, instead of computing the normalized versions inside the `euclidean()` function every time we call `euclidean()`. Try to implement that functionality.
- Similarly, we can make `mpdist.py` faster by storing the normalized versions of the subsequences and using them for the Euclidean distance computations.
- Create an image similar to *Figure 7.4* but for larger time series. Begin with time series with 1,000,000 elements and see what you get.

8
Conclusions and Next Steps

You can either use iSAX as a traditional index or something more sophisticated, as we did in *Chapter 7*. With this practical application of iSAX, we have reached the last chapter of this book! Thank you for reading this book, which is a collaborative effort of lots of people and not just the author.

Time series and time series data mining, in general, are hot topics both in academia and industry, mainly because nowadays, data usually comes in a time series manner. As if this were not enough, those time series contain a lot of data that we need to process quickly and accurately.

This chapter will give you directions on where and what to look at next if you are really into time series or databases.

Disclaimer: All mentioned books and research papers are personal preferences. Your own taste or research interests might vary.

In this chapter, we will cover the following main topics:

- Concluding all that we have learned so far
- Other variations of iSAX
- Interesting research papers on time series
- Interesting research papers on databases
- Useful books

Concluding all that we have learned so far

Time series are everywhere! But time series tend to get larger and larger as we collect more data, more frequently. Therefore, we need ways to process and search large time series faster and faster in order to make useful deductions from the data.

The iSAX index is here to help you search your time series data fast. I hope that this book has given you the necessary tools and knowledge to begin working with time series and subsequences, as well as the iSAX index in Python. However, the knowledge and the presented techniques are easily transferable to other programming languages, including but not limited to *Swift*, Java, C, C++, Ruby, Kotlin, *Go*, *Rust*, and *JavaScript*.

We believe that we have provided the right amount of knowledge about time series indexing using the right amount of theory and practice so you can successfully work with time series and develop iSAX indexes.

The next section presents improved versions of iSAX.

Other variations of iSAX

This book has taught you the initial form of iSAX. There exist more variations of iSAX that make its operation and its construction much faster. However, the core functionality remains the same. You can look into them by reading the following research papers:

- *iSAX 2.0: Indexing and Mining One Billion Time Series*, written by Alessandro Camerra, Themis Palpanas, Jin Shieh, and Eamonn Keogh

- *Beyond one billion time series: Indexing and mining very large time series collections with iSAX2+*, written by Alessandro Camerra, Jin Shieh, Themis Palpanas, Thanawin Rakthanmanon, and Eamonn Keogh

- *DPiSAX: Massively Distributed Partitioned iSAX*, written by Djamel Edine Yagoubi, Reza Akbarinia, Florent Masseglia, and Themis Palpanas

- *Evolution of a Data Series Index: The iSAX Family of Data Series Indexes: iSAX, iSAX2.0, iSAX2+, ADS, ADS+, ADS-Full, ParIS, ParIS+, MESSI, DPiSAX, ULISSE, Coconut-Trie/Tree, Coconut-LSM*, written by Themis Palpanas

Keep in mind that all these are advanced research papers that you might find difficult to understand at first. However, if you are persistent, you will eventually understand them.

In the database area, there exist many indexes because indexing is essential for answering SQL queries fast. Even if they are not directly connected to time series, you might want to have a look at them and adjust them in order to work with time series and subsequences.

A very famous index is called **R-tree**, which is a hierarchical data structure based on the **B+- tree**. You can learn more about the R-tree index by reading *R-trees: A Dynamic Index Structure for Spatial Searching*, written by Antonin Guttman.

Last, a new time series index appeared while writing this book that is also based on the SAX representation and tries to correct the disadvantages of iSAX. The name of the new index is **Dumpy**. The core logic behind Dumpy is the same as in iSAX, but it makes adjustments during the building of the index in order to have a better subsequence distribution inside the index.

You can find more information about Dumpy by reading the paper at `https://arxiv.org/abs/2304.08264`.

The next section mentions some interesting research papers on time series.

Interesting research papers on time series

Here is a list of research papers regarding time series clustering and **anomaly detection** that you might find interesting:

- *A Review on Outlier/Anomaly Detection in Time Series Data*, written by Ane Blazquez-Garcia, Angel Conde, Usue Mori, and Jose A. Lozano

- *Anomaly Detection in Time Series: A Comprehensive Evaluation*, written by Sebastian Schmidl, Phillip Wenig, and Thorsten Papenbrock

- *A Review of Time-Series Anomaly Detection Techniques: A Step to Future Perspectives*, written by Kamran Shaukat, Talha Mahboob Alam, Suhuai Luo, Shakir Shabbir, Ibrahim A. Hameed, Jiaming Li, Syed Konain Abbas, and Umair Javed

- *Time-series clustering – A decade review*, written by Saeed Aghabozorgi, Ali Seyed Shirkhorshidi, and Teh Ying Wah

- *Anomaly Detection for Discrete Sequences: A Survey*, written by Varun Chandola, Arindam Banerjee, and Vipin Kumar

The next section mentions some interesting research papers on databases.

Interesting research papers on databases

As time series are connected to databases, I will give you a small list of classical research papers on databases:

- *Global Query Optimization*, written by Timos K. Sellis

- *The design of POSTGRES*, written by Michael Stonebraker and Lawrence A. Rowe

- *The design and implementation of INGRES*, written by Michael Stonebraker, Gerald Held, Eugene Wong, and Peter Kreps

- *A Relational Model Of Data for Large Shared Data Banks*, written by E. F. Codd
- *The Seattle report on database research*, which can be found at `https://dl.acm.org/doi/10.1145/3524284`

The next section proposes some interesting and valuable books on databases.

Useful books

In this last section of the book, I will list useful books related to computer science and software engineering, starting with the field of databases.

Useful books on databases

As time series are connected to databases, I will give you a small list of classic books on databases:

- *Readings in Database Systems, Fourth Edition*, edited by Joseph M. Hellerstein and Michael Stonebraker
- *Database Internals*, written by Alex Petrov
- *Introduction to Data Mining, 2nd Edition*, written by Pang-Ning Tan, Michael Steinbach, Anuj Karpatne, and Vipin Kumar
- *Database Systems: The Complete Book, 2nd Edition*, written by Hector Garcia-Molina, **Jeffrey D. Ullman**, and Jennifer Widom
- *Database Management Systems, 3rd Edition*, written by Raghu Ramakrishnan and Johannes Gehrke

Databases and time series are not isolated from operating systems and computer programming. Therefore, it would be beneficial to have a strong Computer Science background when working with databases. The next subsection presents some books that will help you with that.

Building a strong computer science background

This subsection presents some books that will help you build a strong computer science background. The list includes the following books:

- *Introduction to Algorithms, 4th Edition*, written by Thomas H. Cormen, Charles E. Leiserson, Ronald L. Rivest, and Clifford Stein
- *Programming Pearls, 2nd Edition*, written by **Jon Bentley**
- *More Programming Pearls: Confessions of a Coder*, written by Jon Bentley
- *Code Complete: A Practical Handbook of Software Construction*, written by **Steve McConnell**
- *Crafting Interpreters*, written by Robert Nystrom

- *The Algorithm Design Manual, 3rd edition*, written by Steven S. Skiena

- *The Elements of Statistical Learning: Data Mining, Inference, and Prediction, Second Edition*, written by Trevor Hastie, Robert Tibshirani, and Jerome Friedman

- *Compilers: Principles, Techniques, and Tools, 2nd Edition*, written by **Alfred Aho**, Jeffrey Ullman, **Ravi Sethi**, and Monica Lam

- *Writing A Compiler In Go*, written by Thorsten Ball

You do not have to read everything from the front cover to the back cover. However, this list of books will give you a strong background in computer science.

The next subsection proposes books that will make you a better UNIX and Linux developer and power user.

Books on UNIX and Linux

This subsection presents some books related to the UNIX and Linux operating systems. The list includes the following books:

- *The UNIX Programming Environment*, written by **Brian W. Kernighan** and Rob Pike

- *The Practice of Programming*, written by Brian W. Kernighan and Rob Pike

- *UNIX Power Tools*, written by Shelley Powers, Jerry Peek, Tim O'Reilly, and Mike Loukides

- *Advanced Programming in the UNIX Programming Environment, Third Edition*, written by **W. Richard Stevens** and Stephen Rago

- *UNIX Network Programming*, written by W. Richard Stevens

- *The C Programming Language, Second Edition*, written by Brian W. Kernighan and **Dennis M. Ritchie**

The next subsection presents handy books related to the Python programming language.

Books on the Python programming language

This subsection presents some books related to the Python programming language. The list includes the following books:

- *Fluent Python*, written by Luciano Ramalho

- *Effective Pandas*, written by Matt Harrison

- *Mastering Python: Write powerful and efficient code using the full range of Python's capabilities, Second Edition*, written by Rick van Hattem

- *Expert Python Programming: Master Python by learning the best coding practices and advanced programming concepts, Fourth Edition*, written by Michal Jaworski and Tarek Ziade

- *Time Series Analysis with Python Cookbook*, written by Tarek A. Atwan.

- *Python Data Analysis, Third Edition*, written by Avinash Navlani, Armando Fandango, and Ivan Idris

- *Machine Learning with PyTorch and Scikit-Learn: Develop machine learning and deep learning models with Python*, written by Sebastian Raschka, Yuxi (Hayden) Liu, and Vahid Mirjalili

With this, we conclude the list of the books you can benefit from.

Summary

In this last chapter of the book, we included a long list of interesting books and research papers. Time series, time series data mining, and databases, in general, are interesting fields that will keep you busy for the rest of your life if you treat them with respect and keep experimenting and learning new things. I can assure you that *you will never get bored* with these fields, either in academia or in industry.

If you retain a single thing from this book, it is to **be curious and experiment all the time**.

However, the first step is finding what interests you the most and following that direction. If you are going to spend a large amount of your time on something, you should definitely find it interesting and challenging enough!

Thank you very much for choosing the book. Feel free to send any suggestions you might have for improving potential future editions of the book. Your comments and suggestions can make a difference!

Useful links

- *Outlier Analysis*, written by Charu C. Aggarwal, Springer, 2013

- The `darts` Python package: `https://pypi.org/project/darts/`

- *R-Trees: Theory and Applications*, written by Yannis Manolopoulos, Alexandros Nanopoulos, Apostolos N. Papadopoulos, and Yannis Theodoridis

- Fluent Python: `https://www.fluentpython.com/`

- Time series data mining using the Matrix Profile part 1: `https://www.youtube.com/watch?v=1ZHW977t070`

- Time series data mining using the Matrix Profile part 2: `https://www.youtube.com/watch?v=LnQneYvg84M`

- Chebyshev polynomials: `https://en.wikipedia.org/wiki/Chebyshev_polynomials`

- The numba Python package: `https://pypi.org/project/numba/`

- iSAX page: `https://www.cs.ucr.edu/~eamonn/iSAX/iSAX.html`

- The Theory of Computation: `https://en.wikipedia.org/wiki/Theory_of_computation`

Exercises

Try to do the following:

- As an exercise, learn about the types of compressed files that are supported by the `pandas.read_csv()` function.

- If you have enough time, try to implement the iSAX index in another programming language such as Go, Swift, or Rust.

- If you are really into Python, you can try optimizing the code of the `isax` package.

- This is a *really difficult* exercise: you can try improving the search performance of the iSAX index by allowing **parallel searching**. Try to implement that functionality in Python with the help of the numba Python package. Personally, I cannot write such a program in Python.

- This is another *really difficult* exercise: try to create a parallel version of the search algorithm that *runs on your GPU*! If you do, please let me know!

Index

<packt>

www.packtpub.com

Subscribe to our online digital library for full access to over 7,000 books and videos, as well as industry leading tools to help you plan your personal development and advance your career. For more information, please visit our website.

Why subscribe?

- Spend less time learning and more time coding with practical eBooks and Videos from over 4,000 industry professionals

- Improve your learning with Skill Plans built especially for you

- Get a free eBook or video every month

- Fully searchable for easy access to vital information

- Copy and paste, print, and bookmark content

Did you know that Packt offers eBook versions of every book published, with PDF and ePub files available? You can upgrade to the eBook version at www.packtpub.com and as a print book customer, you are entitled to a discount on the eBook copy. Get in touch with us at customercare@packtpub.com for more details.

At www.packtpub.com, you can also read a collection of free technical articles, sign up for a range of free newsletters, and receive exclusive discounts and offers on Packt books and eBooks.

Other Books You May Enjoy

If you enjoyed this book, you may be interested in these other books by Packt:

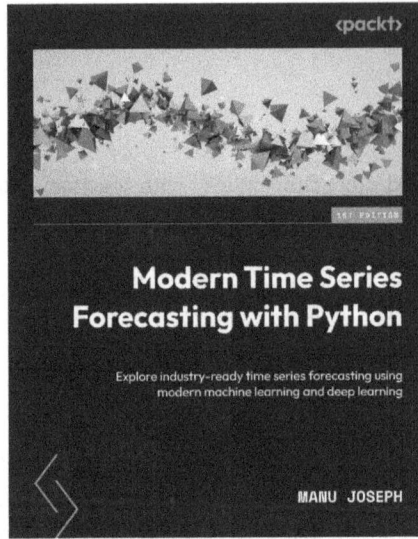

Modern Time Series Forecasting with Python

Manu Joseph

ISBN: 9781803246802

- Find out how to manipulate and visualize time series data like a pro
- Set strong baselines with popular models such as ARIMA
- Discover how time series forecasting can be cast as regression
- Engineer features for machine learning models for forecasting
- Explore the exciting world of ensembling and stacking models
- Get to grips with the global forecasting paradigm
- Understand and apply state-of-the-art DL models such as N-BEATS and Autoformer
- Explore multi-step forecasting and cross-validation strategies

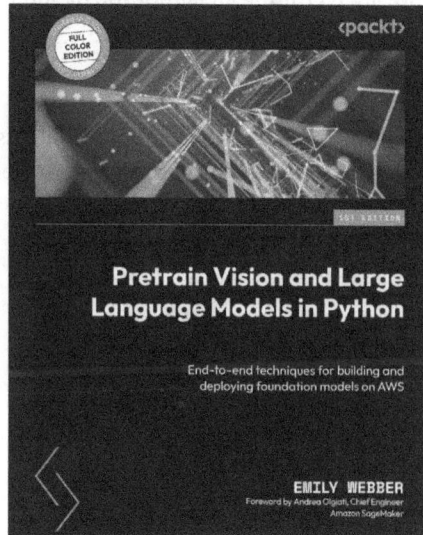

Pretrain Vision and Large Language Models in Python

Emily Webber

ISBN: 9781804618257

- Find the right use cases and datasets for pretraining and fine-tuning
- Prepare for large-scale training with custom accelerators and GPUs
- Configure environments on AWS and SageMaker to maximize performance
- Select hyperparameters based on your model and constraints
- Distribute your model and dataset using many types of parallelism
- Avoid pitfalls with job restarts, intermittent health checks, and more
- Evaluate your model with quantitative and qualitative insights
- Deploy your models with runtime improvements and monitoring pipelines

Packt is searching for authors like you

If you're interested in becoming an author for Packt, please visit `authors.packtpub.com` and apply today. We have worked with thousands of developers and tech professionals, just like you, to help them share their insight with the global tech community. You can make a general application, apply for a specific hot topic that we are recruiting an author for, or submit your own idea.

Share your thoughts

Now you've finished *Time Series Indexing*, we'd love to hear your thoughts! Scan the QR code below to go straight to the Amazon review page for this book and share your feedback or leave a review on the site that you purchased it from.

`https://packt.link/r/1838821953`

Your review is important to us and the tech community and will help us make sure we're delivering excellent quality content.

Download a free PDF copy of this book

Thanks for purchasing this book!

Do you like to read on the go but are unable to carry your print books everywhere?

Is your eBook purchase not compatible with the device of your choice?

Don't worry, now with every Packt book you get a DRM-free PDF version of that book at no cost.

Read anywhere, any place, on any device. Search, copy, and paste code from your favorite technical books directly into your application.

The perks don't stop there, you can get exclusive access to discounts, newsletters, and great free content in your inbox daily

Follow these simple steps to get the benefits:

1. Scan the QR code or visit the link below

https://packt.link/free-ebook/9781838821951

2. Submit your proof of purchase
3. That's it! We'll send your free PDF and other benefits to your email directly